PHOTOCHEMISTRY

PHOTOCHEMISTRY

R. P. WAYNE

AMERICAN ELSEVIER PUBLISHING COMPANY, INC.
NEW YORK

Published in the United States by
American Elsevier Publishing Company, Inc.
52 Vanderbilt Avenue, New York, NY 10017
Library of Congress Catalog Card Number 73-142070
International Standard Book Number 0-444-19600-5

Printed in England

PREFACE

Space research has provided a major stimulus to an interest in the chemical reactions occurring in the atmosphere and beyond. Many of the processes are photochemical in nature, or involve excited species that are most conveniently studied in the laboratory by photochemical techniques. The atoms and molecules concerned in these processes are usually small and simple, but, at the other end of the molecular size scale, much research is directed towards an elucidation of the photochemistry of the complex species that play a part in living organisms, and the results now appearing give considerable insight into the nature of photobiological processes. An appreciation of the advances being made requires some knowledge about fundamental photochemical and photophysical processes.

As a subject, photochemistry has been remarkably poorly represented in terms of published textbooks, and until recently the only available works were at least twenty years out of date. This situation has been remedied at the research level by the appearance of the monumental reference work *Photochemistry* by J. G. Calvert and J. N. Pitts, Jr. (Wiley, New York (1966)). I have felt, however, that there was a need for a much shorter textbook suited especially to the undergraduate or non-specialist research student, and I have written the present book in an attempt to fill this gap. Photochemistry is the natural meeting point for discussions of spectroscopy, energy transfer and the kinetics of fast reactions, and the subject provides a very convenient link between several branches of physical chemistry.

Photochemistry is concerned essentially with the chemistry of excited species, and in this book the subject is developed so as to show the several paths by which an excited species may react or undergo radiative or radiationless decay. Chapters 3–6 describe these paths, and they are preceded by a general introduction to the basic concepts of photochemistry (Chapter 1) and by a brief discussion of the principles of absorption and emission of radiation (Chapter 2). Fairly obviously, there are certain experimental techniques peculiar to photochemistry, and the descriptive material is made more meaningful if the reader understands the nature of the experiments reported. At the same time, the techniques cannot be described until the theoretical background has been developed. I have decided to leave the discussion of techniques until Chapter 7, on the assumption that the reader can refer to that chapter if he wishes to find out about the kinds of method employed for experiments described in earlier chapters. A final chapter (Chapter 8) deals first with some important photochemical processes found in nature and then with certain commercial applications of photochemistry.

This book is intended primarily for non-specialist undergraduate readers, although it is hoped that it may prove useful to graduate students embarking on research in photochemistry. Problems of balance naturally arise in a book of this kind. More detail than an undergraduate can be expected to master must be provided in order to avoid drastic oversimplifications. For example, it would not be justifiable to conceal the complexities of emission phenomena, and Chapters 4 and 5 (together with Sections 2.3 and 2.5, which are relevant mainly to those chapters) contain material that will prove difficult to an undergraduate audience. I suggest, therefore, that an undergraduate should concentrate initially on Chapters 1–3, without spending too much time on Sections 2.3, 2.5 or 3.6. He should also read the introductory sections of Chapters 4–6, together with Sections 4.2, 5.7 and 6.2. The omitted sections should then be meaningful on a second reading. Chapters 7 and 8 do not present any special difficulties. Some knowledge of elementary physical chemistry (e.g. atomic and molecular structure, spectroscopy, reaction kinetics) is assumed.

Specific literature references are not generally given; instead, a bibliography is provided at the end of each chapter to enable the reader to pursue the topics discussed.

I have been most fortunate in having for my own teachers and colleagues some of the world's outstanding photochemists, and I should like to record here my debt of gratitude to my first teacher, Professor R. G. W. Norrish, F.R.S., and to Professor C. H. Bamford, F.R.S., Dr. E. J. Bowen, F.R.S., Professor J. N. Pitts, Jr. and Dr. B. A. Thrush. I wish also to record my appreciation of the advice given me by many other colleagues during the preparation of this book. I am further indebted to Dr. Bowen for reading the first draft of the book and suggesting several changes which I believe have improved the presentation.

Finally, I should like to thank my wife, Brenda, who gave me invaluable assistance with the proof-checking and who helped to make the book more readable, and Mrs. Elizabeth Price, who managed to type the book from a seemingly illegible manuscript.

Oxford,   R. P. Wayne
March 1970

CONTENTS

1

Basic principles of photochemistry

1.1 SCOPE OF PHOTOCHEMISTRY

Man has been aware from the earliest times of the influence that the sun's radiation has on matter; Alexander the Great was one of the first exploiters of a photochemical process when he equipped his Macedonian troops with a treated cloth 'wristwatch' which, under the influence of the sun, changed colour during the day. However, it is during the last fifty years or so that a systematic understanding of photochemical processes has developed. A logical pattern to the interaction between light and matter emerged only after the concept of the quantisation of energy was established. It is the purpose of this book to explain the physical foundations on which modern photochemistry is based; the specific examples given in the book are intended to illustrate these principles rather than to provide a comprehensive survey of known photochemical reactions.

At this stage it would be as well to consider the scope of the rather loosely applied term 'photochemistry'. While an important part of photochemistry is concerned with the chemical change that may be brought about by the absorption of light, a number of physical processes which do not involve any overall chemical change lie within the province of the photochemist; processes such as *fluorescence* (in which light is emitted from a species which has absorbed radiation) or *chemiluminescence* (in which light is emitted as a 'product' of a chemical reaction) must be regarded as of a photochemical nature. The word 'light' is also used loosely, since radiation over a far wider range of wavelengths than the visible spectrum is involved in processes which would be accepted as photochemical. The long wavelength limit is probably in the near infra-red (say at 2000 nm) and the region of interest

1

extends into the vacuum ultra-violet and is limited only formally at the wavelengths where radiation becomes appreciably 'penetrating' (X-rays). No purpose is served in attempting to define more closely the meaning of photochemistry, and the preceding remarks are offered merely to show at the outset the general scope of the subject. The essential feature of photochemistry is probably the way in which 'excited' states of atoms or molecules play a part in the processes of interest. It is apparent that absorption or emission of radiation to or from these states is the concern of the spectroscopist as well as of the photochemist, and the photochemist must have at least a background knowledge of spectroscopy. At the same time, the photochemist is frequently interested in the *rates* at which processes occur, so that the concepts of *reaction kinetics* are often employed. It is assumed that the reader of this book has had contact with the ideas of quantum theory, spectroscopy and reaction kinetics, and that he can obtain access elsewhere to more detailed discussions of these topics than it is possible to provide here.

1.2 LIGHT AND ENERGY

Planck developed his theory of black-body radiation on the basis of a postulate that radiation possessed particulate properties and that the particles, or *photons*, of radiation of specific frequency v had associated with them a fixed energy ε given by the relation

$$\varepsilon = hv \qquad (1.1)$$

where h is called Planck's constant. This quantum theory of radiation was then used by Einstein to interpret the photoelectric effect. As early as the beginning of the nineteenth century, Grotthus and Draper had formulated a law of photochemistry which stated that only the light absorbed by a molecule could produce photochemical change in the molecule. The development of the quantum theory led to a realisation that the radiation would be absorbed in the quantised energy packets; Stark and Einstein suggested that one, and only one, photon was absorbed by a single particle to cause its photochemical reaction. It is now appreciated that several processes may compete with chemical reaction to be the fate of the species excited by absorption (see Section 1.5) and a more satisfactory version of the Stark–Einstein law states that *if a species absorbs radiation, then one particle is excited for each quan-*

tum of radiation absorbed. Although this law might appear trivial in the present-day climate of acceptance of the quantum theory, the law is of fundamental importance in photochemistry, and the agreement between experiment and predictions based on the law does, in fact, offer substantial evidence in favour of the quantum theory of radiation.†

It is now apparent that the energy of excitation of each absorbing particle is the same as the energy of the quantum given by the Planck relation (1.1), and the excitation energy per mole is obtained by multiplying this molecular excitation energy by N, Avogadro's number. A linear relationship exists between energy and frequency, so that frequency characterises radiation in a particularly meaningful way. It has been, however, almost universal practice to discuss the visible and ultra-violet regions of the spectrum in terms of *wavelength* of the radiation, and it is therefore convenient to express the molar excitation energy, E, in terms of wavelength, λ,

$$E = Nhv = \frac{Nhc}{\lambda} \qquad (1.2)$$

where c is the velocity of light. Numerical relationships between E and λ may be derived from the values given for constants in Appendix 1; one particularly useful form is

$$E = \frac{108\,320}{\lambda}\,\text{kJ mol}^{-1} \qquad (1.3)$$

where λ is in nanometres‡ $\left(= \frac{25\,890}{\lambda}\,\text{kcal mol}^{-1} \right)$

Although the chemist frequently finds thermal energy units (kcal) most useful, it is sometimes convenient to express energies

---

†A number of photochemical processes are recognised in which more than one quantum of radiation is absorbed by a single molecule. Many such processes do not violate the Stark–Einstein law: they involve excitation to successively higher energy states of the molecule, each step requiring a single quantum. However, true biphotonic absorption is believed to occur occasionally under conditions of intense illumination; this matter is discussed further in Chapter 5 (p. 154).

‡The unit of wavelength adopted throughout this book is the nanometer (nm) $\equiv$ 10^{-9} m = 10^{-7} cm, in accordance with current practice; the millimicron (mμ) is numerically identical (1μ = 10^{-4} cm). However, many photochemists and spectroscopists prefer to use the Angstrom (Å) $\equiv$ 10^{-8} cm = 10^{-1} nm as the unit of wavelength.

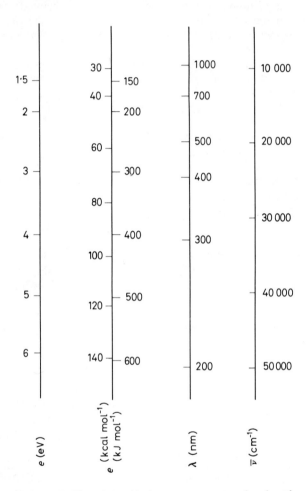

Figure 1.1. The relationship between energy, wavelength and frequency

directly (e.g. in joules, ergs or electron-volts): the conversions between the different units are given in Appendix 2. Figure 1.1 shows the relationships between energy, wavelength and frequency in the visible and near-ultra-violet regions of the spectrum. A useful way of remembering the approximate energies of visible radiation is that the visual range extends from 400 nm to 700 nm, while the corresponding energies are roughly from 70 kcal (293 kJ) mol^{-1} to 40 kcal (167 kJ) mol^{-1}.

1.3 EXCITATION BY ABSORPTION

A molecule which has absorbed a quantum of radiation becomes 'energy-rich' or 'excited' in the absorption process. It is well known by spectroscopists that absorption in the wavelength region of photochemical interest leads to *electronic excitation* of the absorber. Absorption at longer wavelengths usually leads to the excitation of vibrations or rotations of a molecule in its ground electronic state. Although it would be a mistake to suppose that the *only* form of excitation which could lead to photochemical change is electronic, it is generally true that electronically excited states are involved in photochemical processes. The importance of electronic excitation is in part a result of the energy possessed by the 'hot' species, as we shall see in the next section. There is, however, another reason, connected with the nature of the excitation, why electronically excited species exhibit reactivities distinguishable from those of the unexcited species. A simple example will make this clear. The electronic structure of lithium in its ground state is represented in the familiar form Li $1s^2$ $2s^1$: the electrons are placed in the lowest orbitals available to them. The configuration Li $1s^2$ $3p^1$ lies at a higher energy than the $1s^2$ $2s^1$ configuration, and represents an electronically excited state of lithium. Now this excited lithium atom possesses an electron in a p orbital. Since the chemistry of a species depends to a considerable extent on the electronic structure, the reactivity of the excited lithium atom can be expected to differ from that of the ground state atom, quite apart from considerations of the extra energy possessed in the excited configuration. Rather more subtle forms of electronic excitation are possible, and these will be explored further in Chapter 2. For the present purposes it is instructive to take another atom as a further example. Ground state atomic carbon has the electron configuration C $1s^2$ $2s^2$ $2p^2$. Hund's rules of maximum multiplicity tell us that in the ground state the two electrons in the p orbitals are, in fact, unpaired, and therefore occupy two of the three p orbitals. The atom in which two p electrons are paired is a higher energy, excited atom, although the nature of the excitation is now less apparent than in the case of lithium discussed earlier. For any one atomic or molecular species, a great many electronically excited states may be accessible: for each of these states there may be a distinct chemistry which is by no means identical with that of the ground state. Thus it is apparent that the reactions which are observed in photochemical studies may have little in common with the thermal reactions of the parent, ground state, species.

1.4 THERMAL AND PHOTOCHEMICAL REACTIONS

The essential distinction between thermal and photochemical reactions now needs to be explored more fully. Thermal energy may be distributed about all the modes of excitation in a species: in a molecule these modes will include translational, rotational and vibrational excitation as well as electronic excitation. However, for species in thermal equilibrium with their surroundings, the Boltzmann distribution law is obeyed. This law states, of course, that the relative numbers of particles, n_1 and n_2, in two energy levels 1 and 2, separated by an energy gap ΔE, is given by the expression

$$\frac{n_2}{n_1} = e^{-\Delta E/RT} \tag{1.4}$$

If we take a typical energy of an electronically excited state equivalent in thermal units to 60 kcal (251 kJ) mol^{-1}, we can show that at room temperature $(RT \sim 600 \, cal \, (2510J) \, mol^{-1}) \, n_2/n_1 = e^{-100}$ $\sim 4 \times 10^{-46}$, so that a negligible fraction of the species is excited. To achieve a concentration of only 1 % of the excited species would require a temperature of around 6800°C, and in the event most *molecular* species would undergo rapid thermal decomposition from the ground electronic state and it would not be possible to produce appreciable concentrations of electronically excited molecules. In contrast, if molecules absorb radiation at a wavelength of about 500 nm as a result of an electronic transition, then electronic excitation certainly *must* occur, and the concentration produced depends on several factors, including the intensity of illumination and the rate of loss of the excited species. That chemical change can follow the production of electronically excited species is not surprising when consideration is given to the energies involved. The very rough wavelength range suggested in Section 1.1 as being of photochemical interest is similar in equivalent energy to the values of chemical bond energies normally encountered. Thus, if the electronic excitation energy can in some way be made available for bond rupture, then chemical change may occur. The study of photochemistry is concerned in part with the manner and extent of such energy utilisation. Again, the energy of excitation is similar to the activation energies frequently observed for the reaction of unexcited species: if the electronic excitation can be used to overcome all or some of the energy of activation, then it may be expected that the excited species will react *more rapidly* than the ground state species. So we see that photochemical reactions are distinguished from thermal reactions, first by the relatively large concentrations of highly excited species which may react faster than the ground state species, and may even participate isothermally in processes which are endothermic for the latter; and secondly, if the excitation is electronic, by the changes in chemical reactivity which may accompany the new electronic configuration of the species.

A secondary feature of photochemical excitation is that a specific state of the species is formed if the radiation is contained within a sufficiently narrow band: an essentially monoenergetic product can result. It is true that the species may still possess about its excited level an energy-spread characteristic of the temperature of the surroundings, but at room temperatures the range of energies within which most particles lie is very small compared with the energy of excitation, and even narrower distributions may be achieved at reduced temperatures. The possibility of forming monoenergetic species is of particular concern in connection with theories of reaction kinetics, where it is of fundamental interest to see how rapidly a species possessing a specific amount of energy can participate in a reaction. Monoenergetic species can be produced thermally only by the use of sophisticated methods such as the molecular beam technique, while simple photochemical experiments can achieve a relatively narrow energy distribution for electronically excited species.

1.5 FATES OF ELECTRONIC EXCITATION

Photochemical processes involving the absorption of light can be divided into the act of absorption, which falls within the domain of spectroscopy, and the subsequent fate of the electronically excited species formed. It has been implied, in the discussion of the last few pages, that there are several such fates, and we shall now consider more explicitly what the possibilities are. At this stage a highly simplified picture will be presented: each of the processes mentioned will be explored in greater detail later in the book.

Figure 1.2 represents, in simplified form, the various paths by which an electronically excited species may lose its energy.

Energy transfer, represented by paths (iv) and (v) in the diagram, leads to excited species which can then participate in any of the

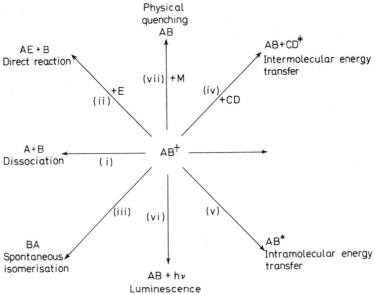

Figure 1.2. The several routes to loss of electronic excitation. The use of the symbols †, * and ‡ is only intended to illustrate the presence of electronic excitation and not necessarily differences in states. One or both of the products of processes (i)–(iii) may be excited

general processes. In this preliminary discussion, therefore, energy transfer will not be considered further; instead this topic is deferred until Chapter 5.

Chemical change can come about either as a result of dissociation of the absorbing molecule into reactive fragments (process i), or as a result of direct reaction of the electronically excited species (process ii); electronically excited species may also undergo spontaneous isomerisation, as indicated by path (iii). Some examples will illustrate these processes: a dagger is used to denote electronic excitation.

$$NO_2\dagger \longrightarrow NO + O \text{ [Dissociation]} \qquad (1.5)$$

$$O_2\dagger + O_3 \longrightarrow 2O_2 + O \text{ [Reaction]} \qquad (1.6)$$

$$cis-C_6H_5CH\!=\!CHC_6H_5 \dagger \begin{cases} trans-C_6H_5CH\!=\!CH\,C_6H_5 \quad (1.7) \\ \text{(Isomerisation)} \\ \\ \end{cases}$$

stilbene

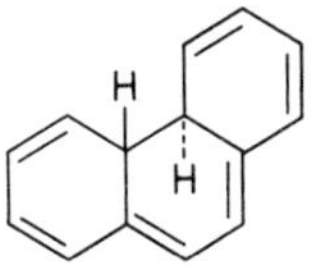

dihydrophenanthrene

Several mechanisms for dissociation are recognised (they include *optical dissociation, predissociation, induced predissociation,* and so on), and they are discussed in more detail in Chapter 3.

Radiative loss of excitation energy (path vi) gives rise to the phenomenon of *luminescence*: the terms *fluorescence* or *phosphorescence* are used to describe particular aspects of the general phenomenon. Luminescence is subject to the laws of radiative processes, and it is more conveniently treated after the discussion of Chapter 2.

The final path (vii) indicated in Fig. 1.2 is *physical quenching.* In this process an atom or molecule M can relieve AB† of its excess energy. Physical quenching differs only formally from intermolecular energy transfer in that M, which must initially take up some excitation energy, does not make its increased energy felt in terms of its chemical behaviour. The electronic excitation of AB† is, in fact, frequently converted to translational or vibrational excitation of M.

1.6   SECONDARY REACTIONS: INTERMEDIATES

We have considered, so far, only the immediate fate of electronic excitation in an absorbing species. It is obvious that the products of processes (i)–(iii) (for example, the oxygen atoms produced in reaction 1.5 or 1.6) can themselves participate in chemical reactions which are, therefore, a direct consequence of the initial absorption of a quantum of radiation. It is usual to differentiate between these *secondary* reactions and the *primary* photophysical processes. As an illustration of these steps, consider the photolysis of ketene (*photolysis* is used to describe a process in which the absorption of light is followed by the rupture of a bond):

$$CH_2CO + h\nu \longrightarrow CH_2CO\dagger \left.\rule{0pt}{2.4em}\right\} \quad \text{(1.8a)}$$

$$\text{Primary process}$$

$$CH_2CO\dagger \longrightarrow CH_2 + CO \quad \text{(1.8b)}$$

$$CH_2 + CH_2CO \longrightarrow C_2H_4 + CO \quad \text{Secondary process (1.9)}$$

Methylene (CH_2) formed in the primary process subsequently reacts with ketene to yield ethylene and carbon monoxide.

Secondary reactions are the ordinary thermal reactions of the various participating species; they are photochemical only in the sense that the reactive species would not have appeared in the absence of light. Certain types of chemical species are found as intermediates more commonly in photochemical than in thermal reaction systems, largely as a result of the energies involved. These intermediates include free atoms and radicals (in which latter class we can include CH_2 produced in reaction 1.8b) as well as electronically excited species. Such intermediates are frequently highly reactive, and their lifetime in a reaction system is correspondingly short. However, reactivity must not be confused with instability; a free radical or atom *in isolation* would have perfectly normal stability, while the lifetime of an electronically excited species would be determined by the probability of losing energy by radiation. Atoms and radicals may themselves carry some kind of excess energy: for example, photolysis of ketene may yield ground state or electronically excited methylene, depending on the wavelength of radiation absorbed.

Chain reactions are very typical of atoms and radicals, and such processes are frequently encountered in photochemistry. An example of a rapid chain reaction is the photochemically initiated hydrogen–chlorine explosion:

$$Cl_2 + h\nu \longrightarrow Cl + Cl \quad \text{(1.10)}$$

$$Cl + H_2 \longrightarrow HCl + H \quad \text{(1.11)}$$

$$H + Cl_2 \longrightarrow HCl + Cl \quad \text{(1.12)}$$

Here reaction (1.10) describes the primary process up to the formation of chlorine atoms; the chain propagation steps are secondary reactions, and both atomic hydrogen and chlorine are *chain carriers*. The ultra-violet photolysis of ozone affords an illustration of a chain reaction involving energy-rich atomic and molecular species:

$$O_3 + h\nu \longrightarrow O_2\dagger + O* \tag{1.13}$$

$$O_2\dagger + O_3 \longrightarrow 2O_2 + O \tag{1.14}$$

$$O* + O_3 \longrightarrow O_2 + O_2\ddagger \tag{1.15}$$

$$O + O_3 \longrightarrow 2O_2 \tag{1.16}$$

$$O_2\ddagger + O_3 \longrightarrow 2O_2 + O* \tag{1.17}$$

The process is of some complexity, and is still not fully understood. It is clear, however, that the reaction intermediates include electronically excited and ground state atomic oxygen as well as at least two different excited molecular oxygen species.

1.7 QUANTUM YIELDS

A concept of great value in photochemistry is that of the *quantum yield, ϕ*. As originally understood, it was *the number of molecules of reactant consumed for each photon of light absorbed*. In this form the quantum yield reflects without distinction both the efficiency of the primary photochemical process in bringing about chemical change, and also the extent of secondary reaction. A quantum yield greater than unity suggests the occurrence of secondary reactions, since the Stark–Einstein law indicates that not more than one molecule can be decomposed in the primary step (a quantum yield greater than two points to the operation of a chain reaction mechanism). However, the discussion of Section 1.5 will have shown that chemical change is not the only consequence of absorption of radiation. Thus a chain reaction may be taking place in a photochemical reaction, even though the overall quantum yield is less than unity. It is more meaningful to consider both primary and overall quantum yields; the primary quantum yield should be stated for a specific primary process (in relation to Fig. 1.2, to one of the paths i–vii). If, for example, dissociation precedes secondary chemical reactions, the primary quantum yield would be the number of molecules dissociating in the primary step for each quantum of light absorbed, and the ratio of overall to primary quantum yields then indicates the extent of secondary reactions. Where nothing to the contrary is stated, overall quantum yield refers to the removal of reactant, although, if several different secondary paths exist, it may be desirable to quote an overall quantum yield for the formation of a specific product.

The determination of overall quantum yields for chemical change requires measurement of the numbers of molecules of reactant consumed, or of product formed, and of the number of quanta of radiation absorbed. The former measurement just involves suitable analytical techniques, while the latter requires a method for measuring absolute numbers of photons. The experimental procedures adopted for such absolute measurements are described in Chapter 7. In the determination of primary quantum yields the contribution to chemical change of secondary reactions must first be eliminated or allowed for, and the absolute efficiencies of radiative and non-radiative energy-loss processes must be assessed. It is not always possible even to establish what primary paths exist, so that a full description of the primary processes in terms of quantum yields can be made only in favourable cases. Nevertheless, several indications may be used to gain some insight into the primary process. The nature of the absorption spectrum may suggest the electronic configuration of the excited state and, hence, the probable fate of the energy. Detection of the intermediates (excited states as well as atoms and radicals) may reveal the products of the primary step (see, for example, Section 7.5). Measurement of overall quantum yields can also give some information about the primary process. If $\phi_{overall} \ll 1$, then in all probability little chemical change occurs in the excited absorbing molecule (although 'cage' recombination of radicals in condensed-phase reaction systems is another very common cause of low quantum yields; cf. Section 3.7). A search must then be made for radiation emitted from the system; the spectrum will indicate whether the emission is fluorescence of the absorber or whether it is derived from a state populated by intermolecular or intramolecular energy transfer. Study of *fluorescence quenching* (Chapter 4) will yield information about physical deactivation processes. Again, if the quantum yield for formation of a specific product is invariant with experimental conditions, such as reactant concentrations or temperature, then that product probably appears, at the measured efficiency, in the primary process.

The energy of an excited species must go somewhere, so the Stark–Einstein law leads to the conclusion that the sum of the quantum yields for *all* primary processes, including deactivation, must be unity. Where sufficient experimental data are available, this expectation is well substantiated.

1.8   REACTION KINETICS

The ratio of overall to primary quantum yields, ϕ_o/ϕ_p, is analogous to the kinetic chain length, v, determined in studies of thermal chain reactions. The quantities may be expressed in terms of rate constants for the several secondary reactions, and their variation with concentrations of various species may lead to confirmation of a hypothetical reaction mechanism and evaluation of rate constants.

In thermal reactions v is defined by the relation

$$v = \frac{\text{rate of reactant disappearance}}{\text{rate of initiation}} \qquad (1.18)$$

Quantum yields may also be defined in terms of *rates*, rather than *numbers* of molecules and photons. An *intensity* of radiation, I, refers to an energy per unit time, and it is frequently convenient to express the *absorbed intensity*, I_{abs}, as the energy absorbed in unit time by unit volume: it is then in the same form as a rate expressed in concentration units, with energies (numbers of photons) replacing numbers of molecules. Hence, for a process

$$A + hv \rightarrow \text{products} \qquad (1.19)$$

$$\phi_o = \frac{-d[A]/dt}{I_{abs}} \qquad (1.20)$$

Furthermore, if we assume that the primary quantum yield, ϕ_p, is for formation of reactive intermediates, then $\phi_p I_{abs}$ is the rate of initiation in the photochemical system, and

$$\frac{\phi_o}{\phi_p} = \frac{\text{rate of reactant disappearance}}{\text{rate of initiation}} \equiv v \qquad (1.21)$$

In fact, initiation by photochemical means is often the best way in which to study the kinetics of radicals or energy-rich species, since not only may the rate of initiation be measured accurately, but also the temperature at which the experiment is performed may be sufficiently low to prevent the occurrence of a plethora of confusing processes often found in thermal radical reactions initiated at high temperatures. A very simple example will show the use of quantum yield measurements in the elucidation of reaction mechanisms and rate constants. The photolysis of ozone–oxygen mixtures by *red* light (where the excited atoms and molecules of reactions 1.13–1.17 are not produced) might be expected to proceed via the following mechanism:

$$O_3 + h\nu_{red} \xrightarrow{\phi_1} O_2 + O \tag{1.22}$$

$$O + O_3 \xrightarrow{k_2} 2O_2 \tag{1.23}$$

$$O + O_2 + M \xrightarrow{k_3} O_3 + M \tag{1.24}†$$

Atomic oxygen is a highly reactive intermediate, and it is possible to apply the *stationary state hypothesis*‡ to solve the rate equations

---

†The 'third body', M, is necessary in reaction (1.24), as in many other atom recombination reactions, to stabilise the vibrationally 'hot' molecule formed immediately after recombination. Oxygen and ozone have differing efficiencies as M; experimental evidence suggests that $k_{3(M=O_2)} = 0.44k_{3(M=O_3)}$; k_3 in the subsequent discussion is taken to be that for O_2 as third body, and [M] is calculated as $([O_3]/0.44 + [O_2])$.

‡'Stationary state' treatments are used frequently in this book; in case the reader is unfamiliar with this method of kinetic analysis, a brief explanation is given here. The concentration of a reaction intermediate is determined, at any instant, by the rates of its formation and loss. For example, consider the processes

$$A + B \xrightarrow{k_4} X \tag{1.28}$$
$$X + C \xrightarrow{k_5} products \tag{1.29}$$

The concentration of the intermediate, [X] is described by the differential equation

$$\frac{d[X]}{dt} = k_4[A][B] - k_5[X][C] \tag{1.30}$$

and, in principle, this equation may be solved explicitly to yield [X] at any time t. However, the kinetic treatment is greatly simplified if it is possible to assume that $d[X]/dt = 0$ (i.e., that the rates of formation and loss are equal so that the intermediate is in a *steady state*), since [X] may be immediately calculated. It is clear that, if [A], [B] and [C] do not themselves alter appreciably, then $d[X]/dt \to 0$ for sufficiently long reaction times. The problem of whether the stationary state hypothesis can be applied is thus one of whether the rate of loss of X approaches the rate of formation in a time short compared both with the time over which reactant concentrations are effectively constant and with the time over which the kinetics of reaction are studied. The applicability of the hypothesis can be tested directly by solving Eq. (1.30) for [X] to find how closely this value approaches the 'stationary state' value. A general result is that the rate of loss of X must be large; that is, the intermediate must be highly reactive. The stationary state analysis is always approximate, although, in situations where it may be used, the errors are usually insignificant.

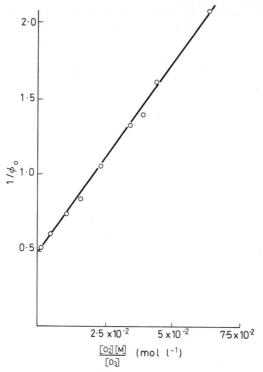

Figure 1.3. Plot of $1/\phi_o$ against $[O_2][M]/[O_3]$ for ozone photolysis by red light. (Data of E. Castellano and H. J. Schumacher, Z. phys. Chem., Frankf. Ausg. 34, 198 (1962))

for $[O]$. Hence, we obtain the relation

$$-\frac{d[O_3]}{dt} = \frac{2\phi_1 I_{abs} k_2 [O_3]}{k_2 [O_3] + k_3 [O_2][M]} \qquad (1.25)$$

or

$$\phi_o = -\frac{d[O_3]/dt}{I_{abs}} = \frac{2\phi_1 k_2 [O_3]}{k_2 [O_3] + k_3 [O_2][M]} \qquad (1.26)$$

Inversion of Eq. (1.26) yields the expression

$$\frac{1}{\phi_o} = \frac{1}{2\phi_1}\left(1 + \frac{k_3 [O_2][M]}{k_2 [O_3]}\right) \qquad (1.27)$$

Figure 1.3 shows $1/\phi_o$ plotted against $[O_2][M]/[O_3]$ for some results on ozone photolysis by red light obtained by Castellano and Schumacher. The graph is sensibly linear over the entire range, which suggests that the hypothetical mechanism may be correct. An intercept ($= 1/2\phi_1$) of very nearly 0·5 is obtained, and the primary quantum yield is therefore close to unity. Primary processes other than decomposition in reaction (1.22) need not be sought. The slope of the graph suggests that, at 18°C, $k_3/k_2 \sim 50\,l$ mol^{-1}: the value is useful confirmation of a similar value obtained in thermal flow experiments (Mathias and Schiff, *Discussion Faraday Soc.* **37**, 38 (1964)).

In this section we have seen how an analysis of reaction rate data can lead to a better understanding of primary and secondary photochemical processes. The kinetic approach is a valuable adjunct to studies of absorption spectra, fluorescence and other optical phenomena, and will be implicit in many of the discussions of the following chapters.

BIBLIOGRAPHY

General Accounts

E. J. BOWEN, *Chemical aspects of light,* Oxford University Press (1946).

G. K. ROLLEFSON and M. BURTON, *Photochemistry and the mechanism of chemical reactions,* Prentice-Hall, New York (1939).

W. A. NOYES and P. A. LEIGHTON, *The photochemistry of gases,* Am. Chem. Soc., New York (1941).

N. J. TURRO, *Molecular photochemistry,* Benjamin, New York (1966).

Standard reference works

J. G. CALVERT and J. N. PITTS, JR., *Photochemistry,* Wiley, New York (1966).

O. L. CHAPMAN, *Organic photochemistry,* Arnold, London (1967).

Articles

P. SUPPAN, 'Photochemistry: some recent advances and current problems', *Chemistry in Britain* **4**, 538 (1968).

J. N. PITTS, JR., F. WILKINSON and G. S. HAMMOND, 'The vocabulary of photochemistry', *Adv. Photochem.* **1**, 1 (1963).

2

Absorption and emission of radiation†

2.1 INTRODUCTION

Photochemistry is intimately dependent on processes involving the absorption or emission of radiation, and it seems desirable to provide at least a brief summary of these processes. Detailed description of spectroscopic phenomena would, however, be out of place, and for amplification of the remarks offered here the reader should refer to one of the texts listed in the Bibliography at the end of the chapter. The survey of this chapter is intended only to serve as a reminder of those parts of spectroscopy needed later in the book.

Spectroscopic nomenclature is often used by photochemists to denote a specific electronic state of some species. Such nomenclature may be the most convenient way of defining a state and distinguishing it from others, although it may sometimes be unnecessarily precise. Those not familiar with the terminology may stop to unravel the meaning of a term symbol when all that had been intended by the writer was to denote that some chemical species was (or was not) excited. A section (2.6) of this chapter sets out briefly the terminology adopted in this book.

2.2 ELECTROMAGNETIC RADIATION

An appreciation of absorption or emission processes requires some understanding of the nature of light. The particular question which we have to ask ourselves as photochemists is how light can

†The formulae and equations of this chapter are written in CGS form, since almost all textbooks on spectroscopy to which the reader may wish to refer use the CGS system.

alter the electronic configuration in an absorbing species, or how a change in the configuration can lead to emission of light.

From the time of Newton until the advent of quantum theory, the corpuscular (or 'particle') theory of light lost ground to the wave theory. Phenomena such as diffraction, or more especially interference, were only explicable in terms of a wave theory. However, the actual nature of the wave, and the mechanism of its propagation, was not established until the latter part of the nineteenth century. In the 1860s, James Clerk Maxwell made one of the major contributions to physics: possibly the only earlier work of such stature was that of Newton. Maxwell was attempting to reconcile the laws of electricity with those of magnetism. By powerful mathematical reasoning, Maxwell demonstrated that such reconciliation would be possible if associated with an oscillating magnetic field there were a similar electric field, and vice versa, and *if a wave were propagated in a direction perpendicular to a plane containing the electric and magnetic fields*. The derivation of Maxwell's equations need not concern us here, but one feature of the equations is of the greatest importance. The velocity of propagation of Maxwell's 'electromagnetic' waves *in vacuo* was shown to be the ratio of electromagnetic to electrostatic units of charge. Purely electrical measurements show this ratio to be numerically identical to the velocity of light *in vacuo*, as determined by Römer (1675), Fizeau (1849) or Foucault (1862). This striking result (1865) obviously suggests that light is an electromagnetic wave, but it did not draw much attention until after Hertz had confirmed (1887–8) Maxwell's prediction of propagated waves from systems involving oscillating electrical and magnetic fields.

The events leading to the awareness that light is a form of electromagnetic radiation have been emphasised here, since a scientist of the second half of the twentieth century has as part of his 'culture' the belief that light is a form of electromagnetic radiation. We also understand that radio waves, infra-red radiation, X-rays and cosmic rays, as well as light and ultra-violet radiation, are electromagnetic radiation, and that they differ one from the other only in terms of their frequencies. The most significant modification of Maxwell's nineteenth century picture of electromagnetic radiation is our awareness that wave motion may have associated with it particulate properties, and that the energy of the particle, or photon, ε, and frequency, v, of the wave are related by $\varepsilon = hv$ (see Section 1.2).

Maxwell's electromagnetic field theory describes radiation in terms of oscillating electric, **E**, and magnetic, **H**, vectors aligned in mutually perpendicular planes, both of which are perpendicular to the direction of propagation of the wave. The oscillations of the two fields may be represented as *sine* functions (since any other function may be constructed by a combination of harmonic *sine* functions with suitable amplitudes and phases), such that

$$\mathbf{E}_y = A \sin 2\pi \left(\frac{x}{\lambda} - vt \right) \tag{2.1}$$

$$\mathbf{H}_z = A \sqrt{\frac{\varepsilon}{\mu}} \sin 2\pi \left(\frac{x}{\lambda} - vt \right) \tag{2.2}†$$

The x Cartesian coordinate is here the direction of propagation, so that the electric vector is contained in the xy plane and the

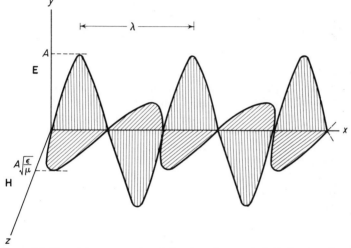

Figure 2.1. Pictorial representation of electric and magnetic fields due to a plane-polarised electromagnetic wave. The figure shows the fields as a function of distance in the direction of propagation, x, at any fixed time

magnetic vector in the xz plane. The *intensity* of radiation is proportional to the square of the electric vector amplitude, A; the amplitude of the magnetic vector is $A\sqrt{(\varepsilon/\mu)}$, where ε, μ are the

†These equations refer to *plane-polarised* light, where $\mathbf{E}_z = \mathbf{H}_y = 0$. Unpolarised light has random orientations of **E** (perpendicular to **H**) vectors; in circularly or elliptically polarised light there is a specific fixed phase relationship between coherent $\mathbf{E}_y$ and $\mathbf{E}_z$ vectors.

dielectric constant and magnetic permeability of the medium through which the wave is propagated ($\varepsilon = \mu = 1$ *in vacuo*). Figure 2.1 is a pictorial representation of Eqs. (2.1) and (2.2) at any instant.

Equations (2.1) and (2.2) indicate that at any point in space (fixed x), electric and magnetic fields oscillate sinusoidally with time at a *frequency* v, while at any instant (fixed t), the amplitudes of the fields vary sinusoidally with distance along the direction of propagation so that the distance between two maxima is the *wavelength*, λ. The product $v\lambda$ is the velocity of propagation of the wave, v. It may be shown that v is related to the velocity of propagation *in vacuo*, c, by the equation

$$v = c/\sqrt{\varepsilon\mu} \qquad (2.3)$$

2.3   ABSORPTION AND EMISSION PROCESSES

In this section an attempt is made to describe absorption and emission processes in terms of mechanical models. A more complete and satisfactory representation of the processes is given mathematically by solutions of the *time-dependent Schrödinger equation*. Presentation of the mathematical equipment needed to deal with this approach would, however, obscure the basic principles which we wish to develop; where necessary, the results of the wave-mechanical reasoning are given without proof. (For a brief introduction to the solutions of the time-dependent Schrödinger equation, see Chapter 4 of G. M. Barrow's book listed in the Bibliography for the present chapter.)

There are three processes which we must distinguish: *absorption, stimulated emission* and *spontaneous emission*. Suppose a chemical species possesses two quantized states l and m, of energies ε_l and ε_m. If the species is in state l initially, it might be able to interact in some way with electromagnetic radiation and absorb energy in order to reach state m. In a normal process this absorption of energy occurs in a single step so that the energy difference between final and initial levels must be equivalent to the energy of a single photon of radiation. Hence, absorption of radiation can only occur if $\varepsilon_m - \varepsilon_l = hv$ ('*Bohr condition*'). The process of *absorption* has involved the loss of intensity of the electromagnetic radiation and the gain of energy by the absorbing species. The converse process, in which a species in an upper state gives up energy to electromagnetic radiation and increases the intensity, is known as *stimulated emission*: the word 'stimulated' indicates that it is the interaction between the radiation already present and the energy-rich species that encourages the latter to yield up its energy. Although we have not mentioned the nature or magnitude of the interaction between the species and radiation, it is apparent that the rate (intensity) of absorption or stimulated emission is proportional to the rate of 'collision' between photons and the absorber or emitter: that is to say, the intensity change is proportional to \the radiation density, ρ, and to the concentration of chemical species. The constant of proportionality defines the so-called Einstein '*B*' coefficients. B_{lm} is the coefficient for absorption, while B_{ml} is that for stimulated emission: the principle of microscopic reversibility suggests that $B_{lm} = B_{ml}$, and this result can also be derived from the complete treatment of radiation theory. The rates of absorption and stimulated emission are $B_{lm}n_l\rho$ and $B_{ml}n_m\rho (= B_{lm}n_m\rho)$, respectively, where n_l and n_m are the concentrations of species in lower and upper states. For a system in thermal equilibrium n_m is always less than n_l (see the Boltzmann equation 1.4, on p. 6), and absorption is always a more important process than stimulated emission. How much more important depends, of course, on the relation between $(\varepsilon_m - \varepsilon_l)$ and the temperature, T. It has been pointed out that the energy levels of significance in photochemistry are such that $(\varepsilon_m - \varepsilon_l) \gg kT$, and $n_m \ll n_l$, so that stimulated emission is rarely important in photochemical processes in which thermal equilibrium is established. However, in non-equilibrium situations, stimulated emission may not be negligible, and if a *population inversion* $(n_m > n_l)$ arises, then the emission process will predominate over absorption, and net emission will result. The LASER (Light Amplification by Stimulated Emission of Radiation) depends on the achievement of such population inversions, generally by photochemical techniques (see Section 5.6).

In addition to absorption and stimulated emission, a third process, *spontaneous emission*, is required in the theory of radiation. In this process, an excited species may lose energy spontaneously — that is, in the absence of a radiation field — to reach a lower energy state. Spontaneous emission is a random process, and the rate of loss of excited species by spontaneous emission (from a statistically large number of excited species) is kinetically first-order. A first-order rate constant may therefore be used to describe the intensity of spontaneous emission: this constant is the Einstein '*A*' factor,

A_{ml}, which corresponds for the spontaneous process to the second-order B constant of the induced processes. The rate of spontaneous emission is equal to $A_{ml}n_m$, and intensities of spontaneous emission can be used to calculate n_m if A_{ml} is known. Most of the emission phenomena with which we are concerned in photochemistry — fluorescence, phosphorescence and chemiluminescence — are normally spontaneous, and the descriptive adjective will be dropped henceforth. Where emission is induced the fact will be stated.

We referred in the last paragraph to the calculation of concentrations of excited species from emission intensity measurements. It may, however, not always be possible to determine A_{ml} directly by the techniques to be discussed in Chapter 7, and some other method of evaluating A factors may be needed. We shall show now that the A coefficient may be calculated from the B coefficient for the same transition (B may be determined from experimental absorption measurements, as will be seen in Section 5 of this chapter).

Consider an assembly of species, capable of undergoing transitions from states l to m and vice versa, contained at thermal equilibrium within a black-body enclosure. The *net* rate of absorption of radiation is

$$B_{lm}n_l\rho - B_{ml}n_m\rho - A_{ml}n_m \qquad (2.4)$$

but, in a system at equilibrium with its surroundings, this must be zero. The Boltzmann distribution for thermal equilibrium gives the relative numbers of molecules in l and m states:

$$n_m/n_l = e^{-(\varepsilon_m - \varepsilon_l)/kT} = e^{-h\nu/kT} \qquad (2.5)$$

Furthermore, ρ (which is a function of frequency, ν) is well known for a black-body enclosure, and is, in fact, embodied in Planck's law:

$$\rho = \frac{8\pi h\nu^3}{c^3} \cdot \frac{1}{e^{h\nu/kT} - 1} \qquad (2.6)$$

Substitution of Eqs. (2.5) and (2.6) into Eq. (2.4), with Eq. (2.4) set equal to zero and with $B_{lm} = B_{ml}$, leads to the relation

$$A_{ml} = \frac{8\pi h\nu^3}{c^3} B_{lm} \qquad (2.7)$$

This equation is the 'ν^3 law' to which reference is frequently made by spectroscopists and photochemists.

The nature of the interaction between electromagnetic radiation and matter must now be considered. The processes may become clearer if we consider a simple example: the absorption of infrared radiation by a molecule of HCl. The molecule has a permanent dipole moment, so that the energy of the molecule will be affected by the presence of an electric field, and the bond will tend to be distorted according to the direction of the field. Now consider an oscillating electric field, such as that present in electromagnetic radiation. If the frequency of oscillation is equal to the vibration frequency of the H—Cl bond, then the induced motion of the electrons may lead to an increased energy of nuclear motion. The vibrational energy in the molecule will then increase by one quantum, and the intensity of electromagnetic radiation will be depleted by an equivalent amount. (The basic tenet of the quantum theory is that the distortions occur *only* in the quantised units which will lead to absorption of a whole photon.) This description of the absorption process is obviously just a pictorial representation, but it indicates that the interaction derives from the influence, via the molecular dipole, of the electric vector of the radiation on the energy of the molecule. A transition occurring through such an interaction is called an *electric dipole transition*. Interactions with the magnetic vector of the radiation, or those which involve quadrupoles in the chemical species, give rise to *magnetic dipole*, *electric quadrupole* and *magnetic quadrupole* transitions, and so on. All these interactions are, however, much weaker than the electric dipole interaction, and may frequently, but by no means always, be ignored.

The absorption or emission of infra-red radiation by an oscillating molecule possessing a dipole is readily understood in the pictorial terms of the last paragraph. It is less easy to describe electronic transitions in the same manner. In the classical sense, electronic excitation does not correspond to increasing the energy in an oscillating system, and, in any case, neither upper nor lower electronic state may possess a steady dipole (e.g. the electron cloud in an atom is symmetrically disposed about the nucleus in all states, so there is no effective charge separation). However, the general principles of interaction with radiation still apply, and what we need to know is whether an (electric) dipole interaction can occur during transitions between two states. Wave-mechanical techniques provide the only satisfactory method of dealing with this problem: the time-dependent Schrödinger equation, referred to at the beginning of the section, can be used to derive the rate at which a system can be changed from one stationary state to another under the influence of a perturbing effect. If this rate is non-zero for a perturbation of the system involving electric dipole interaction with the electric vector of radiation, then an electric dipole transition can occur. The rate of change between states multiplied by the number of species present in the lower state is, of course, the overall rate of absorption of photons, so that, in principle, solution of the time-dependent Schrödinger equation should lead to prediction of the intensity of the transition. Explicit solutions are, however, rarely available, and in such cases it may be possible only to say whether or not any interaction occurs, rather than to calculate its magnitude. The conditions under which a specific interaction arises are given as *selection rules* for that type of transition: for electric dipole transitions in electronic spectroscopy they are given in Section 7 of this chapter. One quantity which appears in the solution to the quantum-mechanical equations must be mentioned — the *transition moment*, $|\mu_{lm}|$ (the alternative symbol R_{lm} is sometimes used). The time-dependent perturbation of energy in, say, the x direction in our system depends upon the interaction between the electric field, E_x, and a dipole moment component, μ_x, and it is found that the equation for the rate of change in the system contains the integral

$$\int_{-\infty}^{+\infty} \psi_m^* \mu_x \psi_l \, dx \qquad (2.8)$$

(ψ_l is the time-independent wave function for state l, and ψ_m^* is the complex conjugate of the wave function for state m). For convenience of handling, the integral (2.8) is represented by the symbol $|\mu_{xlm}|$: the corresponding quantity for isotropic radiation in three dimensions is then the transition moment $|\mu_{lm}|$. The actual result from the time-dependent Schrödinger equation is that the rate of change in the system is

$$\frac{8\pi^3}{3h^2} |\mu_{lm}|^2 \rho \qquad (2.9)$$

where ρ is the radiation density. Earlier in the section we defined a constant of proportionality, B_{lm}, for absorption processes, such that the overall rate of absorption was $B_{lm} n_l \rho$ for n_l species in state l. Comparison with Eq. (2.9) for the rate of change for a single absorber shows that

$$B_{lm} = \frac{8\pi^3}{3h^2} |\mu_{lm}|^2 \tag{2.10}$$

In this section we have distinguished between spontaneous and induced transitions, and we have shown how the probabilities for these processes, the Einstein A and B coefficients, are related both to each other and to a theoretically calculable quantity, the transition moment, $|\mu_{lm}|$. The following section deals with experimental measurements of absorption, and the relation between these measurements and the theoretical quantities is explored in Section 2.5.

2.4 ABSORPTION OF RADIATION: THE BEER–LAMBERT LAW

The fraction of light transmitted through an absorbing system is very frequently found to be represented by the relation

$$\frac{I_t}{I_0} = 10^{-\varepsilon Cd} \tag{2.11}$$

I_t, I_0 are transmitted and incident light intensities, C is the concentration of absorber, and d is the depth of absorber through which the light beam has passed. ε is a constant of proportionality known as the *extinction coefficient*: it is dependent on the wavelength of radiation (and may occasionally vary with C – this question is mentioned later). The law embodied in Eq. (2.11) was originally known as Lambert's law; a second law, Beer's law, stated that if C and d were altered but the product Cd was constant, then the fraction of light transmitted remained the same. Since this latter law follows in any case from Lambert's law, the relation (2.11) is now known as the *Beer–Lambert law*. The logarithmic form of the equation,

$$\log_{10} I_0/I_t = \varepsilon Cd \tag{2.12}$$

is often employed, and the product εCd is called the *Optical Density* (O.D.) of the system.

A 'proof' of the Beer–Lambert law may be derived if it is assumed that the rate of loss of photons is proportional to the rate of bimolecular collisions between photons and the absorbing species.

The decrease, $-dI$, in intensity I at any point x in the system (see Fig. 2.2) for a small increase in x, dx, is given by

$$-dI = \alpha I C dx \qquad (2.13)$$

where α is a constant of proportionality. Integration, with the boundary conditions $I = I_0$ at $x = 0$, $I = I_t$ at $x = d$, yields the result

$$\frac{I_t}{I_0} = e^{-\alpha C d} \qquad (2.14)$$

which is the same as Eq. (2.11) with $\alpha = 2 \cdot 303\varepsilon$.

Both natural and decadic extinction coefficients are used in practice, and it is essential to state the base as well as the units of

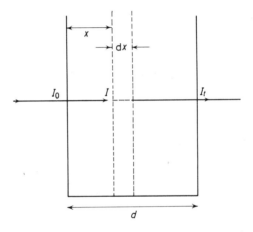

Figure 2.2. The change in intensity, I, with optical path, x (see Eqs. 2.13 and 2.14)

C in referring to extinction coefficients (the units of d are almost always cm). Thus a 'decadic molar extinction coefficient' is ε of Eq. (2.11) with C in mol l^{-1} and d in cm.

The intensity of radiation absorbed, I_{abs}, is, of course, $I_0 - I_t$, so that the fraction absorbed is given by

$$\frac{I_{abs}}{I_0} = 1 - 10^{-\varepsilon C d} = 1 - e^{-\alpha C d} \qquad (2.15)$$

An important approximate expression results when αCd is small: expansion of the exponential $e^{-\alpha Cd}$ and rejection of second- and higher-order terms in αCd leads to the conclusion that

$$\frac{I_{abs}}{I_0} \sim \alpha Cd \qquad (2.16)$$

For $\alpha Cd = 0{\cdot}01$, the approximate value of I_{abs}/I_0 differs from the accurate one by less than 1 %, while even for $\alpha Cd = 0{\cdot}1$, the difference is only 5 %. Thus Eq. (2.16) often gives a sufficiently accurate estimate of fractional absorptions less than about 10 %.

More than one absorber may be present in some absorbing systems. The rate of bimolecular collisions between photons and each species is dependent on the intensity and the concentration of each species, so that the right-hand side of Eq. (2.13) must have terms for the other components added on: integration will now yield the result

$$\frac{I_t}{I_0} = e^{-(\alpha_1 C_1 d + \alpha_2 C_2 d \dots)} = 10^{-(\varepsilon_1 C_1 d + \varepsilon_2 C_2 d \dots)} \qquad (2.17)$$

where the subscripts refer to the different components.

Although the Beer–Lambert law usually offers an adequate description of experimental data, there are some circumstances in which it does not. For example, the width of an absorption band or line depends in part on factors such as molecular collision, so that changes in concentration can alter the ε–λ relationship and hence lead to a breakdown of the law. For oriented systems (e.g. crystals) the value of ε may depend on the plane of polarisation of the light. Again, the law holds only if the wavelength range to which the intensity measurements refer is small compared with the width of the absorption band (i.e. if ε is constant over the wavelength range). Thus experiments in which there is absorption from wide-band incident radiation by a narrow-band absorber will not obey the Beer–Lambert law. More trivially, the concentrations of species which associate or dissociate will not be equal to the concentrations expected on the basis of amounts of material added.

As we shall see later in the book, the maximum value achieved in an absorption band by the extinction coefficient may be an indicator of the nature of the spectroscopic transition, and especially of whether the transition is 'allowed' or 'forbidden' for electric dipole interactions. In the next section it is shown that it is the total absorption throughout a band that is related to the theoretical transition moment, and it follows that the extinction coefficient and the band width bear some kind of inverse relationship for a given value of $|\mu_{lm}|$. However, for a band of typical width the maximum value of the decadic molar extinction coefficient, ε_{max}, for a permitted transition rarely exceeds 10^5: a value of 5×10^4 is considered usual.

2.5   RELATION BETWEEN EXPERIMENTAL AND THEORETICAL QUANTITIES†

A calculation of, say, $|\mu_{lm}|$ from wave functions may be used to provide an estimate of the extinction coefficient at any wavelength. Conversely, the experimental extinction coefficient may be used to gain some information about the wave functions in upper and lower states, or, perhaps, to calculate the Einstein A factor for spontaneous emission. In this section we examine how the experimental and theoretical quantities are related.

Although the experimental extinction coefficients α or ε are measured, in principle, at a single wavelength or frequency, an absorption band may occupy a range of frequencies, and the calculated transition moment, $|\mu_{lm}|$, refers to the whole band. The quantities we wish to compare are, therefore, $|\mu_{lm}|$ and the *integrated absorption coefficient for the band*, $\mathscr{A}$. For the present purpose, we shall define $\mathscr{A}$ by the relation

$$\mathscr{A} = \int \alpha \, dv \qquad (2.18)$$

where the limits of integration are set to be the frequency limits of the transition under consideration. Equation (2.13) may be modified to show the decrease in intensity, $-d\mathscr{I}$, over the whole band:

$$-d\mathscr{I} = \mathscr{A} \, \mathscr{I} C' \, dx \qquad (2.19)$$

The units of concentration are most conveniently taken as molecules per cubic centimetre, and the prime is placed on C to indicate the use of a specific unit. In Section 2.3, it was shown that the rate of change of state in an absorbing system was given by

$$\frac{8\pi^3}{3h^2} |\mu_{lm}|^2 \rho \qquad (2.9)$$

---

†In the interests of simplicity, it is assumed in this section that electronic states are non-degenerate, and that the refractive index of the medium is near unity.

The number of molecules in a cross-sectional area of one square centimetre and a depth dx, is, of course, $C' dx$, so that the rate of removal of photons from the incident radiation is

$$\frac{8\pi^3}{3h^2} |\mu_{lm}|^2 \rho C' \, dx \tag{2.20}$$

Since one photon at the frequency v is equivalent to an energy hv, and since the rate of removal of energy across unit area is $-d\mathscr{I}$, it follows that

$$-d\mathscr{I} = \frac{8\pi^3}{3h^2} |\mu_{lm}|^2 \rho hvC' \, dx \tag{2.21}$$

The intensity $\mathscr{I}$ is an energy flux and is clearly equal to the energy density multiplied by the velocity of propagation, so we may write

$$\rho = \mathscr{I}/c \tag{2.22}$$

and substitute for ρ in Eq. (2.21); after cancellation of h, the result

$$-d\mathscr{I} = \frac{8\pi^3}{3hc} |\mu_{lm}|^2 v \, . \, \mathscr{I}C' \, dx \tag{2.23}$$

is obtained. Comparison of Eq. (2.19) and (2.23) shows that

$$\mathscr{A} = \frac{8\pi^3}{3hc} |\mu_{lm}|^2 v \tag{2.24}$$

Thus we have an expression which relates the experimental quantity $\mathscr{A}$ with the theoretical quantity $|\mu_{lm}|$; Eqs. (2.10) and (2.7) can be used to relate B_{lm} and A_{lm} to $\mathscr{A}$.

It is appropriate at this stage to introduce the concept of *oscillator strength, f,* sometimes used by spectroscopists and photochemists. The oscillator strength arises in classical dispersion theory in the expression for molar refraction, but for present purposes it may be defined as the ratio of the observed $\mathscr{A}$ to a particular reference value. For electronic transitions the reference value is taken to be the value of $\mathscr{A}$ predicted on the assumption that a single electron undergoes the transitions, and that it is attracted to the centre of a spherically symmetric molecule by a force which follows Hooke's law. The wave functions for the electron in various levels (e.g. 0, 1, etc.) are those for a harmonic oscillator, and may be used to derive the result, here stated without proof, that

$$|\mu_{01}|^2 = \frac{3he^2}{8\pi^2 mv} \qquad (2.25)$$

(where e, m are the charge and mass of an electron). Substitution of this result into Eq. (2.24) yields the reference value for $\mathscr{A}$,

$$\mathscr{A}_{ref} = \frac{\pi e^2}{mc} \qquad (2.26)$$

so that

$$f = \frac{mc}{\pi e^2}\mathscr{A} = \frac{8\pi^2 mv}{3he^2}|\mu_{lm}|^2 \qquad (2.27)$$

For many electronic absorption bands, f may be near unity (for example, for the yellow atomic sodium lines familiar in emission); if f is small, the band is likely to correspond to a 'forbidden' transition. As noted at the end of the last section, for many absorption bands of typical width, $\varepsilon_{max} \sim 5 \times 10^4$ for $f \sim 1$.

2.6 SPECTROSCOPIC NOMENCLATURE

The spectroscopic nomenclature used in later parts of the book needs to be explained, although it is out of place to show how, for example, term symbols for atoms are derived. It seems more pertinent to the present purpose to explain the function of the nomenclature, and then to cite briefly the actual forms used in the book.

For the spectroscopist the object of the nomenclature is to define the electronic arrangement — or electronic 'state' — of the two levels involved in a spectroscopic transition; for the photochemist the object may be to specify the excited state implicated in a photo-chemical process. Statement of the principal quantum number of the shell containing the electron in a hydrogen atom would define the electronic arrangement with considerable precision, but for more complex species a more extensive description of the electronic state may be required. Electronic states involving a single electron in an unclosed shell *may* be adequately described by the list of shells and the number of electrons. For example, the ground state of the Na atom may be represented as $1s^2 2s^2 2p^6 3s^1$ and the first electronically excited state as $1s^2 2s^2 2p^6 3p^1$, and the photochemist who wishes merely to know the extent and type of excitation might be satisfied by these descriptions. However, transitions between these 'two' levels give rise to the well-known *doublet* yellow D lines; as a result of differing interactions between electron spin and orbital momentum, there are two different states, of different energies, although both are described by $1s^2 2s^2 2p^6 3p^1$. If, then, a closer definition of the state is required, some further information must be provided. This information is contained within the spectroscopic *term symbol*. The two states corresponding to the description $1s^2 2s^2 2p^6 3p^1$ are given the term symbols $^2P_{\frac{1}{2}}$, $^2P_{\frac{3}{2}}$ (or, since both these term symbols can arise from *any* atom having one electron in a p orbital, the principal quantum number may also be stated, and the term symbols written $3(^2P_{\frac{1}{2}})$, $3(^2P_{\frac{3}{2}})$). It is the meaning of the components of the term symbol with which we are primarily concerned. In atomic term symbols such as $^2P_{\frac{1}{2}}$, the superscript number, subscript number and the capital letter each defines a particular angular momentum of the electronic state, as will be described in the next paragraph. However, it is frequently the *symmetry* properties of two wave functions that determine whether or not the electric dipole transition moment vanishes, so that from the spectroscopist's point of view symmetry and energy may be the most important features of an electronic state. As it happens, the angular momentum quantum numbers provide some of the symmetry information; and for atoms or for some molecules, a statement of angular momentum quantum numbers forms at least part of the spectroscopic nomenclature. We now proceed to the nomenclature used for certain types of species: atoms will be discussed in greatest detail as the principles are simplest to understand.

Atoms

Let us consider first a single-electron atom (that is, atomic hydrogen or an atom with closed inner shells which has only one 'valence' electron). The Bohr model of the atom pictures the electron orbiting the nucleus, and we also know that the electron spins on its own axis; the electron therefore possesses both *orbital angular momentum* and *spin angular momentum,* which are vector quantities and are given the symbols **l** and **s**. Orbital and spin momenta combine to produce a *total angular momentum,* **j**, according to the laws of vector addition, but with the added restriction for quantised momenta that the addition can take place only in quantised units. What this means is that for a single electron, where $s = \frac{1}{2}$, j cannot take all values between $l+\frac{1}{2}$ and $l-\frac{1}{2}$, but only those two extreme values. This combination of angular momenta is called *coupling*. When there are several electrons in an atom, coupling between the angular momenta of the electrons can take place to give angular momenta for the atom. It is these resultant angular momenta that are described in the atomic term symbol. We are concerned here with *Russell–Saunders coupling*, in which the individual orbital momenta l_1, l_2, etc., couple to give an overall orbital momentum, **L**, the spin momenta s_1, s_2, etc., couple to give **S**, and **L** and **S** then couple to give an overall total angular momentum **J**. Such coupling describes the situation fairly adequately for light atoms, but not for heavy ones; however, Russell–Saunders term symbols are often used for heavy atoms and the breakdown of selection rules based on **L**, **S** and **J** (see Section 2.7) reflects the inadequacy of the description.† Once again the vector additions of **l** and of **s**, and finally of **L** and **S** to give **J**, must take into account the quantisation of the angular momenta. Thus J can take values $L+S$, $L+S-1, L+S-2, \ldots, L-S$ if $L > S$, and there are $(2S+1)$ values of J. $(2S+1)$ is said to be the *multiplicity* of the state, and is the superscript which precedes the capital letter in the term symbol, while the magnitude of J is the subscript which follows it. The capital letter itself represents the value of L, so formally the term symbol is

$$^{2S+1}L_J$$

In order to avoid writing out the number for L, we represent $L = 0, 1, 2, 3$, etc., by the capital letters S, P, D, F, etc., following directly the convention for individual orbital notation that electrons with $l = 0, 1, 2, 3$ should be called s, p, d, f electrons.

Although it was agreed at the beginning of this section not to give details of the derivation of term symbols, the remarks of the last paragraph might be clearer if a simple example were given.

---

†Pictorially the situation can be seen as the decreasing tendency of the different **l** (or **s**) vectors to couple to each other to make **L** (or **S**) and an increasing tendency for the individual **l** and **s** vectors to produce individual **j** vectors, as the electrons become further apart from each other. For heavy atoms '*j–j*' coupling (in which individual **j** vectors couple to make the final resultant **J**) may therefore be a more faithful representation than Russell–Saunders coupling.

The sodium atom has only one electron of interest, and $L = l$, $S = s$ and $J = j$. In the first excited state $(\ldots 3p^1)$, $l = 1$, $s = \frac{1}{2}$, and, therefore, $j = \frac{1}{2}$ or $\frac{3}{2}$. Hence, the states are $^2P_{\frac{1}{2}}$, $^2P_{\frac{3}{2}}$, as given earlier.† In speech, the states are called 'doublet-P-one-half' and 'doublet-P-three-halves', the states with multiplicities 1, 2, 3, etc. originally being known as singlets, doublets, triplets, etc. after the appearance of spectra involving them.

Diatomic and linear polyatomic molecules

In diatomic and linear polyatomic molecules the electronic state may still be defined in part by the magnitude of the orbital electronic angular momentum resolved along the internuclear axis. The nomenclature follows the general pattern, with Roman letters replaced by Greek characters. Thus l and L for atoms are replaced by λ and Λ for individual orbitals and the whole molecule, respectively, and orbitals with $\lambda = 0, 1, 2, 3$ are the familiar σ, π, δ, φ orbitals, while states with $\Lambda = 0, 1, 2, 3$ are Σ, Π, Δ, Φ states. The term symbol consists basically of

$$^{2S+1}\Lambda$$

although a number of other pieces of information may be added on. One of these is, of course, the total angular momentum, and one of several possible coupling schemes must be used to derive it. Secondly, some further description of the symmetry properties of the wave function may be possible. A wave function may possess some, or all, of the molecular symmetry. In particular, for a centrosymmetric molecule, the wave function may either remain unchanged or change sign (but not magnitude) on inversion through the centre of symmetry. Such wave functions are called 'even' or 'odd', respectively: in German the words are 'gerade' and 'ungerade' and the symbols g or u, given as subscripts after Λ, derive from the initial letters. The wave function for a Σ state $(\Lambda = 0)$ may remain the same or change sign on reflection by a plane of symmetry passing through the line of atomic centres: these two possibilities are represented by the symbols $+$ or $-$ appearing as superscripts after Λ. A few examples will illustrate the forms used.

---

†For the ground state, with one s electron, $S = \frac{1}{2}$, $L = 0$ and, since J must be positive, it can only take the value $\frac{1}{2}$. In general, where $L < S$, the multiplicity is $2L+1$, and the values of J are $S+L$, $S+L-1, \ldots, S-L$. For the ground state of sodium, therefore, the term symbol is $^2S_{\frac{1}{2}}$.

N_2 – the ground state is $^1\Sigma_g^+$: there is neither spin nor orbital angular momentum since all orbitals are closed (fully occupied), and the wave function does not change sign either on inversion through a centre of symmetry or on reflection.

CO_2 – the ground state is $^1\Sigma_g^+$: the same remarks apply to this linear triatomic molecule as to N_2.

N_2O – the ground state is $^1\Sigma^+$: since the molecule has no centre of symmetry, the wave function cannot be either g or u.

NO – the ground state is $^2\Pi$: there is one electron in an un-filled orbital, which is an anti-bonding π orbital. S is therefore $\frac{1}{2}$, $(2S+1) = 2$; Λ is 1, and the molecule is in a Π state. NO does not possess a centre of symmetry.

O_2 – the ground state is $^3\Sigma_g^-$: there are two electrons in the antibonding π orbitals which can give rise to $S = 0$ or 1 and $\Lambda = 0$, 1 or 2. Not all the apparent combinations are allowed because of the operation of the exclusion principle. $^3\Sigma_g^-$ is the lowest energy arrangement, while the states $^1\Delta_g$ and $^1\Sigma_g^+$ (among others) are of rather higher energy. Note that several energy levels can derive from different coupling of electrons, without moving an electron to a different orbital.

Small, non-linear molecules

It is not practicable to specify the electronic state of a non-linear molecule in terms of orbital angular momentum, although spin multiplicity is still meaningful. If the molecule possesses symmetry elements, and the electrons are still sufficiently delocalised for the electronic cloud to possess effectively the same symmetry, then it may be possible to describe the electronic state in terms of the effect that the symmetry operations have on the sign of the wave function. The behaviour of the wave function under the different operations is most conveniently expressed as the *group-theoretical symmetry type* for the particular symmetry group. Symbols such as A, B, E and T, with numerical subscripts after them, refer to these sym-metry types. Group theory is a powerful tool for handling sym-metry problems of the kind involved here, but cannot be discussed any further at the present time. Suffice it to say that a wave function of type A_1 is unaltered by any symmetry operation, one of type A_2 will have its sign altered by certain operations, and so on. The numerical subscript may be followed by g or u if the molecule is centrosymmetric. Examples of the nomenclature for some simple molecules are given below:

	Ground state	Typical excited state
$\begin{array}{c}H \\ \diagdown \\ C=O \\ \diagup \\ H\end{array}$:	$^1A_1$	$^1B_2$
(benzene ring) :	$^1A_{1g}$	$^1E_{1u}$
NO_2 :	$^2A_1$	$^2B_1$

Complex molecules

Complex molecules may not possess any symmetry elements, or if they do, the localisations of the electrons can so distort the electron cloud that its symmetry bears little relation to the molecular symmetry. In such cases it may be best to revert to a description of states in terms of the individual orbitals. As an example, we will consider formaldehyde, although a molecule as simple as this is probably best described by the group-theoretical term symbol of the last paragraph. The last filled orbitals in H_2CO can easily be shown to be ... $(\pi_{CO})^2 (n_O)^2$, where n_O represents the non-bonding orbital on the O atom and the two electrons in it are the 'lone pair'. Since all orbitals are filled, the state is a singlet (because all electron spins must be paired), and as the lowest-lying singlet it is sometimes represented as S_0: this kind of nomenclature is frequently met in the photochemistry of organic molecules. The first excited singlet is called S_1, and so on; excited triplets are T_1, T_2, etc., and vibrational excitation is shown by a superscript v (e.g. S_1^v). The first unfilled orbitals in formaldehyde are the π_{CO}^* and σ_{CO}^* anti-bonding orbitals. Promotion of one electron from, say, the n_O orbital to π_{CO}^* leads to excitation, and there is no restriction now on unpairing of spins so both singlet and triplet states are possible. The states are then designated as $^3(n, \pi^*)$ or $T_1(n, \pi^*)$, $^1(n, \pi^*)$ or $S_1(n, \pi^*)$, and similarly for (π, π^*), (n, σ^*) and (π, σ^*) states. Spectroscopic transitions are then referred to as $n \rightarrow \pi^*$, and so on.

In conclusion, it must be said that several other notations exist — for example, one devised by Mulliken — but in this book we will use only those described.

2.7   SELECTION RULES FOR OPTICAL TRANSITIONS

Formal rules, known as *selection rules*, may be used to decide whether or not an electric dipole transition between two states may take place. Since these rules will be alluded to at several points later in the book, it seems desirable to list the more important at this stage.

Perhaps the most important is the rule governing spin multiplicity: spin must not change during an electronic transition. The usual way to write rules of this kind is

$$\Delta S = 0 \qquad (2.28)$$

In atoms, for one-electron transitions, we have the selection rules

$$\Delta L = \pm 1 \qquad (2.29)$$

$$\Delta J = 0, \pm 1, \text{ but } J = 0 \not\rightarrow J = 0 \qquad (2.30)$$

For diatomic and linear polyatomic molecules the orbital momentum rule is

$$\Delta \Lambda = 0, \pm 1 \qquad (2.31)$$

and, where applicable, the rules governing symmetry

$$\text{u} \leftrightarrow \text{g}, \; + \leftrightarrow +, \; - \leftrightarrow - \qquad (2.32)$$

Selection 'rules' may also be derived for the symmetrical small molecules discussed in Section 2.6 under 'Small, non-linear molecules'; the basis for these rules is that $|\mu_{lm}|$ will vanish unless $\psi_m^* \mu \psi_l$ is totally symmetric (i.e. of type A_1). The wave function for the ground state, ψ_l, is frequently of type A_1, and it can be shown that $\psi_m^* \mu \psi_l$ will be of type A_1 if ψ_m^* and any of μ_x, μ_y or μ_z are of the *same* symmetry type. Hence, in any given case, transitions can be divided into 'allowed' and 'forbidden'.

Absorption or emission spectra due to 'forbidden' transitions are occasionally observed, and although they are often much weaker than 'allowed' transitions, there are some instances where the transition probability is high. One such example, which is of the greatest importance in photochemistry, concerns mercury 'resonance' radiation. (Resonance radiation refers to emission from the first excited state to the ground state: *resonant* absorption and re-emission occurs, and light may escape only from the edges of the emitting body.) The ground state of mercury is $^1S_0$, and the first excited singlet is $^1P_1$. Transitions of the $^1P_1 \rightarrow {}^1S_0$ line yield the true resonance line at 1849 Å, and it is so readily re-absorbed by mercury vapour that it is necessary to cool mercury lamps used for its generation. The first excited *triplet* states are $^3P_{0,1,2}$ and transitions of the $^3P_1 \rightarrow {}^1S_0$ line at $\lambda = 253.7$ nm are only weaker by a factor of about a hundred than those of the $\lambda = 184.9$ nm line, even though they are nominally forbidden by the $\Delta S = 0$ selection rule. The intensity of the forbidden line is, in fact, so great that the line at $\lambda = 253.7$ nm is commonly called the resonance line. The explanation for the breakdown of the rule in this instance is that, since mercury is a heavy atom, Russell–Saunders coupling does not really hold. Thus **S** is not a good quantum number for mercury, and selection rules based on it need not be expected to apply rigorously.

Three frequent causes of apparent breakdown of selection rules are listed below.

(1) Transition may be forbidden for electric dipole interaction, but permitted for magnetic dipole or quadrupole interactions, etc.

(2) The quantum number to which the selection rule refers may be inapplicable, as in the example given above.

(3) Collisions with other atoms or molecules, or the presence of electric or magnetic fields, may invalidate selection rules based on state descriptions of the unperturbed species.

2.8 INTENSITIES OF VIBRATIONAL BANDS IN AN ELECTRONIC SPECTRUM

Relative intensities of different vibrational components of an electronic transition are of importance in connection with both absorption and emission processes. The populations of the vibrational levels obviously affect the relative intensities. In addition, electronic transitions between given vibrational levels in upper and lower states have a specific probability determined in part by the electronic transition probability, and in part by the probability of finding a molecule with similar internuclear separations in both states. This last factor is bound up with the *Franck–Condon principle*. The principle states that an electronic transition occurs so rapidly in comparison with vibration frequencies that no change in internuclear separation occurs during the course of a transition. Thus a line depicting a transition on a potential energy diagram must be drawn vertically. The probability of transition occurring in any small range of internuclear distance will then depend on the product of probabilities of a molecule possessing that internuclear distance in each electronic state, and the total transition probability is this probability integrated over all internuclear distances.

Quantum-mechanical treatment of oscillating motion in molecules shows that the internuclear distance–probability function

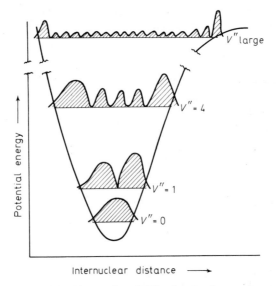

Figure 2.3. Vibrational probability function for a series of levels of an anharmonic oscillator

depends on the quantum number v, and that the function has $(v+1)$ maxima and has v nodes between the maxima. Figure 2.3 shows the probability function for a series of levels of a typical anharmonic oscillator: the greater heights of the maxima for large internuclear distance at high values of v correspond classically to the greater time spent by the molecule at the turning point where the restoring force is smaller. The relative intensity of transitions can now be predicted by the Franck–Condon principle. In absorption, transitions are most likely to originate from the point of maximum probability in the particular vibrational level of the lower state of the transition, and the relative intensities of transitions to vibrational levels of the upper state will be dependent

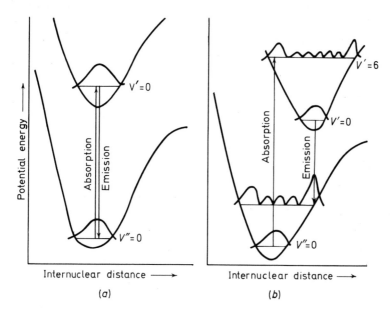

Figure 2.4. Electronic transitions of greatest probability for absorption and for emission from lowest vibrational levels (a) where both electronic states are of similar sizes, (b) where the upper state is larger than the lower

on the probability of finding the upper state with that internuclear separation. Figure 2.4 illustrates the two different situations that arise (a) when the upper and lower curves are similar in shape and size, and (b) when the upper state is larger than the lower state. In case (a) it can be seen that (0,0) transitions are the strongest, in both absorption and emission, while in case (b) the strongest absorption band is (6,0) and the strongest emission is (0,4). For transitions not involving $v'' = 0$ (absorption) or $v' = 0$ (emission), the $(1,1)$, $(2,2)$, . . . , etc., bands are strong in case (a), and $(5,1)$, $(4,2)$, . . . , etc., are strong in case (b).

BIBLIOGRAPHY

G. M. BARROW, *Introduction to molecular spectroscopy:* Chapter 4, 'The absorption and emission of radiation'. McGraw-Hill, New York (1962).

G. HERZBERG, *Spectra of diatomic molecules,* 2nd edn, Van Nostrand, Princeton, N.J. (1950).

G. HERZBERG, *Electronic structure and electronic spectra of polyatomic molecules,* Van Nostrand, Princeton, N.J. (1966).

D. H. WHIFFEN, *Spectroscopy,* Longmans, London (1966).

3

Photodissociation

3.1 DISSOCIATION AS A PRIMARY PROCESS

An examination of primary photochemical processes may well begin with a discussion of photodissociation, since, of the possible fates of electronically excited species, dissociation most clearly leads to chemical change. The several different paths indicated in Fig. 1.2 are, however, closely interrelated, and in this chapter we shall have to anticipate certain conclusions about fluorescence and energy transfer which are developed more fully in later chapters of the book.

We shall distinguish three major routes to photodissociation — optical dissociation, predissociation and induced predissociation — and the physical principles involved are described in Sections 3.2–3.4. The illustrative examples chosen in these sections generally involve diatomic inorganic molecules. The potential energy–internuclear distance relationship for a diatomic species is represented in two dimensions by the familiar potential energy curve, and the actual form of the curve is frequently known for specific electronic states of diatomic molecules. Although the same physical principles will apply to the photochemistry of larger molecules, the descriptions are necessarily more complex and less precise. Later sections of the chapter deal with polyatomic molecules of both organic and inorganic species.

3.2 OPTICAL DISSOCIATION

An electronically excited species produced by the absorption of light may possess enough, or more than enough, energy to dissociate into fragments. The spectrum of the absorption leading to dissociation is continuous, since the fragments may possess translational energy (which is effectively continuous). At some longer wavelength the spectrum may possibly be banded (although in some cases it is not), in a region where dissociation does not follow absorption. The spectrum of I_2 vapour exhibits typical absorption bands. The bands get progressively closer together until a continuum is reached: the energy corresponding to the onset of the continuum ('convergence limit') is the dissociation energy to the products formed. At room temperature, almost all iodine molecules are in the ground vibrational level ($v'' = 0$),

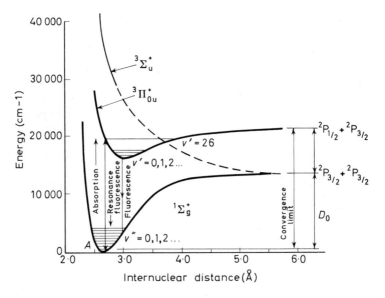

*Figure 3.1. Potential energy curves for the ground state and two excited states of the I_2 molecule. (After L. Mathieson and A. L. G. Rees, J. chem. Phys. **25**, 753 (1956))*

so that the bands are a progression from $v'' = 0$ and the dissociation energy is the energy from that level. Figure 3.1 gives approximate potential energy curves for the states of I_2 involved (see Section 3.4 for further remarks). The electronic transition $^3\Pi \leftarrow {}^1\Sigma$, although 'forbidden' by the selection rule $\Delta S = 0$, is fairly strong for the heavy molecule I_2 (it becomes progressively weaker for Br_2, Cl_2 and F_2). The potential energy curve indicates that the upper state of I_2 *correlates* (lies on the same curve as) $I(^2P_{\frac{1}{2}})+$

$I(^2P_{\frac{1}{2}})$: that is, photodissociation yields one excited ($J = \frac{1}{2}$) as well as one ground state ($J = \frac{3}{2}$) product atom. One or more excited product fragments are often produced in optical dissociation, and comparison of the spectroscopic convergence limit with the known bond dissociation energy may well reveal the nature of product excitation. For example, in iodine the continuum starts at $\lambda = 498.9$ nm, equivalent to 88.0 kcal (368 kJ) mol^{-1}, while the I—I bond dissociation energy is 36.1 kcal (151 kJ) mol^{-1}. The difference of 51.9 kcal (217 kJ) mol^{-1} corresponds exactly to the excitation energy of the $J = \frac{3}{2}$ to $J = \frac{1}{2}$ state of atomic iodine.

Determination of the precise convergence limit by direct examination of spectra is difficult, and sometimes impossible; a more reliable estimate of the dissociation limit is usually obtained from an extrapolation of measurements taken well into the banded region. Analytical methods may be used occasionally (especially if the potential curve is well represented by the Morse function), but the graphical *Birge–Sponer extrapolation* is of more general application. A comprehensive discussion of the methods is given by Gaydon (Bibliography).

Although the convergence of absorption bands to a continuum is typical of the spectra of small molecules which are optically dissociated by light absorbed in the continuum, there are some molecules for which the spectrum is continuous over the entire absorption region. A continuous spectrum suggests that either or both of the electronic states involved in the transition are unbound, or so weakly bound that the spacing of vibrational levels is too small to be resolved. It is, of course, improbable that the *lower* state of an ordinary molecule is so weakly bound that it leads to a continuous spectrum, even though some 'quasi' molecules, such as Hg_2 formed in high-pressure mercury vapour, give rise to an absorption continuum. Many molecules do, however, possess repulsive excited states. Figure 3.2 shows some states of hydrogen iodide which correlate with ground state $H(^2S_{\frac{1}{2}})$ and $I(^2P_{\frac{3}{2}})$ atoms; a broad absorption band extends from $\lambda \sim 300$ nm to $\lambda < 180$ nm, and results in part from the $^3\Pi \leftarrow {}^1\Sigma^+$ and $^1\Pi \leftarrow {}^1\Sigma^+$ transitions, and photodissociation occurs at all wavelengths in the band.† Indeed, since the H—I dissociation energy is about 71 kcal (297 kJ) mol^{-1}, there is always considerable excess energy ($\lambda = 300$

---

† A transition to a higher triplet level, which correlates with $H(^2S_{\frac{1}{2}}) + I^*(^2P_{\frac{1}{2}})$, may contribute some absorption to the band. It is not clear what fraction of electronically excited I atoms is formed.

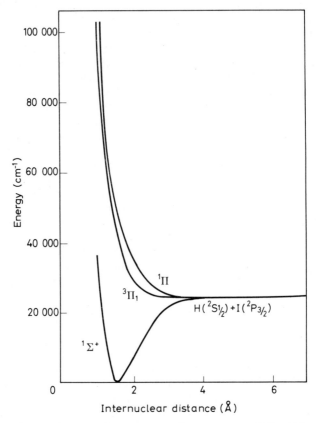

Figure 3.2. Potential energy curves for some states of HI *which correlate with ground state atoms. (After R. S. Mulliken,* Phys. Rev. **51**, *310 (1937))*

nm $\equiv$ 95 kcal (397 kJ) mol^{-1}) which must be taken up as translation of the product atoms, and there is evidence of increased
chemical reactivity of these 'hot' atoms.

Photodetachment of an electron, *photoionisation*, can be regarded
as a special case of photodissociation, with the ion and electron
as dissociation products:

$$AB + hv \rightarrow AB^* \rightarrow AB^+ + e \qquad (3.1)$$

Rydberg series (in which the principal quantum number increases)
are known for both atoms and molecules, and the lines or bands
converge as the electron is moved into orbitals further away from
the nucleus. The convergence limit corresponds to complete removal of the electron, and thus to ionisation. Experiments using
reaction vessels with windows are limited to wavelengths longer
than about 105 nm (the cut-off of lithium fluoride: cf. footnote
on p. 194); photoionisation phenomena have not, therefore,
generally been studied by the photochemist since ionisation
energies often correspond to $\lambda < 105$ nm. The processes are, however, of the greatest importance in the upper atmosphere, where
short wavelength ultra-violet radiation from the sun can lead to
appreciable ionisation of the chemical species present. There are,
nevertheless, a few substances whose ionisation potential is
lower than the lithium fluoride cut-off, of which, perhaps, the
most investigated is nitric oxide. At wavelengths shorter than
134.3 nm absorption of light is followed by ionisation:

$$NO + hv \rightarrow NO^+ + \varepsilon \qquad (3.2)$$

Photoionisation of some *excited* molecules, for which ionisation
of the ground state does not occur at $\lambda > 105$ nm, has also been
observed, and it may be used to characterise and estimate the
excited species. For example, electronically excited O_2 (in the
$^1\Delta_g$ state) may be photoionised by argon resonance radiation,
which is just transmitted by an LiF window.

3.3   PREDISSOCIATION

Complex organic molecules do not usually undergo optical dissociation in the regions of strongest absorption. The increased
number of electronic states, the closer spacing between them and

the large number of vibrational modes all tend to increase the probability of radiationless transitions between states. Thus an excited state populated below its dissociation limit may undergo a radiationless transition to populate another state *above* its dissociation limit. The process is one which, for small molecules, is treated as predissociation, and in this section we describe the physical principles.

The word *predissociation* was adopted to describe the spectroscopic appearance of an absorption system. The gas-phase absorption spectra of the simpler molecules show considerable sharp

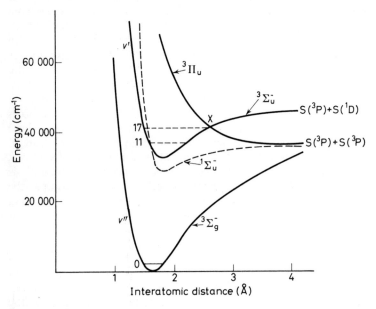

Figure 3.3. Potential energy curves for S_2 *showing the crossings which lead to 'normal' and 'induced' predissociations. (From E. J. Bowen,* Chemical aspects of light, *p. 95, Oxford University Press (1946))*

rotational structure, but in some cases this rotational structure becomes blurred, leading to a diffuseness of the bands at a wavelength longer than that corresponding to the optical dissociation limit. For example, in the absorption spectrum of S_2 there is a region of diffuseness near the 17, 0 band, although the bands are sharp at other wavelengths. (Another region of diffuseness is observed near the 10, 0 band, and will be discussed in Section 3.4.)

Predissociation is now understood to arise from the 'crossing' of electronic states, and the occurrence of radiationless intra-molecular energy transfer between them. Figure 3.3 shows the crossing curves for the S_2 molecule. The diffuseness arises from an increase in the line widths of the individual rotational transitions, and in order to explain it we must first consider the line width of a spectroscopic transition for which there is no radiationless transition. In the absence of molecular collisions, a species will remain in an excited state for an average time (the radiative lifetime, τ_0) of about $1/A$, where A is the Einstein factor for spontaneous emission; the radiative lifetime for a state from which there is an allowed transition to the ground state is of the order of 10^{-8} s. Now the Heisenberg uncertainty principle postulates that it is not possible to know precisely both position and momentum of a particle, and it can also be expressed in terms of the uncertainty ΔE in the energy of the particle and the uncertainty Δt in the time for which it possesses the energy:

$$\Delta E \,.\, \Delta t \approx h/2\pi \qquad (3.3)$$

Since $E = h\nu$, for a spectroscopic transition we may write

$$\Delta \nu \,.\, \Delta t \approx 1/2\pi \qquad (3.4)$$

However, the uncertainty Δt is the same as the radiative lifetime τ_0, so we may write

$$\Delta \nu \approx 1/2\pi\tau_0 \qquad (3.5)$$

Thus any spectral line has an uncertainty in its frequency, $\Delta \nu$, determined by τ_0, and this uncertainty is equivalent to a finite width — the *natural line width* — of the line. In terms of wavelength we may write (since $\nu = c/\lambda$; $d\nu = -cd\lambda/\lambda^2$)

$$-\Delta\lambda = \lambda^2/2\pi c\tau_0 \qquad (3.6)$$

For our permitted transition with $\tau_0 \sim 10^{-8}$ s, this width is therefore about 5×10^{-6} nm at $\lambda \sim 300$ nm, which is much less than the spacing between rotational lines. If, however, Δt becomes less than τ_0, the line width will increase. A decrease in lifetime of an excited state can be brought about by the occurrence of radiation-less transitions to a different state. Radiationless transitions taking place during the period of a few vibrations reduce Δt to about 10^{-13} s, and $\Delta\lambda$ becomes about 0·5 nm, which is now *greater* than the rotational spacing: the spectrum appears diffuse.

The discussion of the last paragraph will have made it clear that for diffuseness of rotational fine structure to be observed, the radiationless transition must occur rapidly enough to give sufficient line-broadening. However, even if radiationless transition occurs only at ten times the rate of spontaneous emission (typically, $A < 10^8 \text{ s}^{-1}$), most of the molecules in the first excited state will pass over to the second state, although there will be no apparent diffuseness in the absorption spectrum. Emission bands will, on the other hand, be drastically reduced in intensity under such circumstances, since most molecules will not survive in the first state long enough to radiate. Thus the 'breaking-off' of bands in emission is a more sensitive test of radiationless transition than is diffuseness in absorption; the term 'predissociation' refers to the transition processes leading to dissociation rather than to the diffuseness in absorption spectra.

Chemical dissociation of the absorber may follow radiationless transition if the transition occurs at an energy sufficient to cause dissociation of the new state. This energy may well be less than the dissociation energy in the state which is populated initially by absorption of radiation. For a diatomic molecule the fragments from photodissociation must be chemically identical whatever the dissociation mechanism. It follows that the state of excitation of fragments produced by predissociation below the optical dissociation limit must be lower than those formed by optical dissociation in a continuous absorption region. The solid potential energy curves for S_2 (Fig. 3.3) show that predissociation yields two ground state atoms ($S(^3P)$), while optical dissociation would lead to one excited, $S(^1D)$, atom. It is important to note that, while predissociation may lead to the formation of some product at wavelengths longer than the dissociation limit, the products may well not be identical with those from optical dissociation.

The Franck–Condon principle (Section 2.8) applies to radiationless, as well as to radiative, transitions, and the nuclei do not move significantly during the course of the change in electronic state. If two potential curves intersect, then the two states possess the same total (electronic + vibrational) energy at the same internuclear distance. (If the curves do *not* intersect, either the internuclear distance must change (a violation of the Franck–Condon principle) or kinetic energy must be released instantaneously.) Radiationless transition occurs, therefore, at the internuclear distance, and at the energy, represented by the crossing point of two intersecting potential energy curves (e.g., the point X in Fig. 3.3). The actual rate of crossing is determined in part by whether the radiationless transition is 'allowed' or not, and in part by the overlap of the vibrational probability curves at and near the crossing point (cf. Section 2.8). Selection rules have been derived for radiationless transitions, analogous to those for optical transitions. They are, where applicable,

$$\Delta S = 0 \tag{3.7}$$

$$\Delta J = 0 \tag{3.8}$$

$$\Delta \Lambda = 0, \pm 1 \tag{3.9}$$

$$+ \leftrightarrow + \qquad g \leftrightarrow g$$
$$\tag{3.10}$$
$$- \leftrightarrow - \qquad u \leftrightarrow u$$

Except for the rule $g \leftrightarrow g$, $u \leftrightarrow u$, the rules are identical with the optical transition selection rules (2.28), (2.31) and (2.32).

Another factor which determines the probability of crossing is the length of time spent by the molecule in the configuration of the crossing point. This is determined by the velocity of approach, which is a measure of the excess energy possessed by the molecule over that required for crossing. It is a frequent observation that a spectrum which shows diffuseness as a result of predissociation becomes sharp again at shorter wavelengths.

Three subdivisions of the predissociation mechanism may be distinguished, and they are represented by the curves of Fig. 3.4. Absorption takes place from state G to a level A of state E_1; crossing occurs at the point X to state E_2. The state E_2 in both cases (a) and (b) is stable, and dissociation of E_2 cannot take place unless the level A in state E_1 is higher than the dissociation limit in E_2, D. Thus the energy corresponding to the onset of diffuseness (the *predissociation* limit) is the *same* as D. (If crossing in case (b) occurs *below* D, the lifetime of E_1 will not necessarily be shortened, since the state may be repopulated by the reverse crossing.) An unstable state E_2 is represented by case (c). The predissociation limit lies *above* the dissociation limit D; at points above D but below X, the transition would necessitate an appreciable change in internuclear distance.

The curves shown for the S_2 molecule in Fig. 3.3 indicate that the predissociation at $\lambda \sim 261 \cdot 5$ nm ($v' \sim 17$) is of type (c). The S_2 bond dissociation energy is 101 kcal (423 kJ) mol^{-1}. Nitrogen dioxide provides an interesting example of predissociation according to case (b) (see also Section 4.3). A region of diffuseness in the absorption spectrum sets in at $\lambda \sim 370$ nm, and is probably a result of predissociation of the $^2B_2$ state, produced on absorption, by an upper $^2A_1$ state which correlates with ground state products (Fig. 3.5). The primary quantum yield for O atom formation have been measured, and are indicated in Fig. 3.6 for a temperature of 24°C. The quantum yield for the photodissociation

$$NO_2 + h\nu \rightarrow NO + O \qquad (3.11)$$

shows a marked rise at wavelengths around the predissociation limit, and at $\lambda < 370$ nm ϕ_0 is probably unity. It is, however, noteworthy that at $\lambda = 404\cdot7$ nm (equivalent to 70·6 kcal (295 kJ) mol^{-1}), $\phi_0 = 0\cdot36$, since the dissociation energy of NO_2 to ground

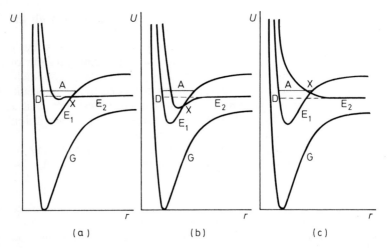

Figure 3.4. Three possible cases for crossings of potential energy curves which can lead to predissociation (see text). (Relabelled from G. Herzberg, Spectra of diatomic molecules, 2nd edn, p. 422, Van Nostrand, Princeton, N. J. (1950))

state products is at least 71·8 kcal (300 kJ) mol^{-1}. It is believed that the additional 1·2 kcal (5 kJ) mol^{-1} can be made up from the vibrational and rotational energy of the NO_2 (especially since it can be shown that rotation can be converted to vibration in the bond-dissociating mode). The temperature dependence of ϕ_0 for $\lambda = 404\cdot7$ nm is strong evidence in favour of this hypothesis. Table 3.1 shows that ϕ_0 increases with temperature, and in a most

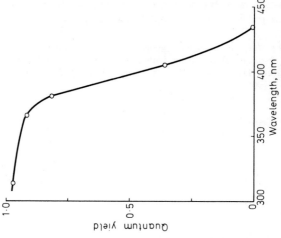

Figure 3.6. *Primary quantum yields for O atom formation in the photolysis of* NO_2 *at various wavelengths. (Data of J. N. Pitts, Jr., J. H. Sharp and S. I. Chan, J. chem. Phys.* **40**, *3655 (1964))*

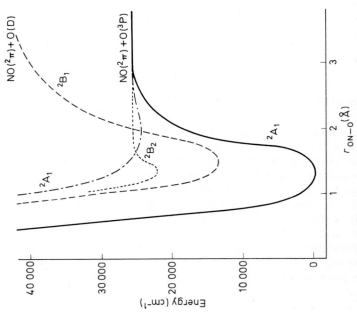

Figure 3.5. *Potential energy curves for some states of* NO_2. *(After J. N. Pitts, Jr., J. H. Sharp and S. I. Chan, J. chem. Phys.* **40**, *3655 (1964))*

convincing way the fractions of molecules possessing more than 1.2 kcal (5 kJ) mol^{-1} of excess energy parallel the values of ϕ_0.

Table 3.1. Variation of primary quantum yield for NO_2 photolysis at $\lambda = 404.7$ nm and of fractions of molecules with $E > 1.2$ kcal (5 kJ) mol^{-2} as a function of temperature. Data of J. N. Pitts, Jr., J. H. Sharp and S. I. Chan, *J. chem. Phys.* **40**, 3655 (1964).

Temperature (°C)	$\phi_0(\lambda = 404.7)$	Fraction of molecules with $E > 1.2$ kcal (5 kJ) mol^{-1} in rotational and vibrational excitation
23	0.36	0.36
133	0.51	0.50
223	0.71	0.70
293	0.91	0.90

The possible variation of primary quantum yield with temperature should always be considered where the temperature-dependence of overall quantum yield is being used to investigate *secondary* reactions. Another example of such variation (for the induced predissociation of Br_2) is given in the next section. Optical dissociation into a continuum may also be important only at elevated temperatures: increased population of $v'' > 0$ may permit absorption to the dissociation continuum at wavelengths for which absorption from $v'' = 0$ is still banded.

3.4   INDUCED PREDISSOCIATION

The efficiency of crossing between two electronic states may be so low that predissociation no longer reduces the intensity of emission bands. Even in the absence of loss processes such as physical quenching, radiative loss will now ensure that most of the excited species do not undergo chemical change. Such inefficiency of intramolecular energy transfer between crossing states usually reflects some degree of 'forbiddenness' of the radiationless transition. There are, however, some cases where the efficiency of the radiationless transition depends on the external environment. We have seen (end of Section 2.7) that collisions with other species, or the presence of magnetic or electric fields, can invalidate the selection rules for *optical* transitions. Similar apparent breakdown of radiationless transition selection rules is observed — the rules are applicable only to unperturbed molecules. The increased probability of crossing between the appropriate states can lead to an increase in the relative extent of predissociation as a molecule becomes more perturbed by its external environment. Predissociation that is significant only in the presence of some perturbation is known as *induced predissociation*.

Collision-induced predissociation may be recognised by the broadening, caused by increased diffuseness, of bands in the absorption spectrum, seen when the pressure is increased or when a foreign gas is added to the absorbing system. Predissociations induced by the absorbing gas itself lead to deviations from the Beer–Lambert law: the apparent absorption increases faster than the law predicts. A similar apparent increase in absorption intensity is seen for predissociation induced by a foreign gas. (The sharper the lines in an absorption band, the weaker the band appears in a low-dispersion spectrum which does not resolve the individual lines, since the unabsorbed background contributes more to the total transmitted intensity.) It should perhaps be mentioned that it may be difficult to distinguish ordinary pressure-broadening of a spectral line from broadening caused by collision-induced predissociation. No such difficulty exists with the quenching of *specific* bands of an emission system in the presence of a foreign gas, or at high pressures.

One of the first instances in which induced predissociation was recognised was in the I_2 molecule. We have already described the banded appearance of the absorption spectrum in terms of the $^3\Pi \leftarrow {}^1\Sigma$ transition (Fig. 3.1). At very low pressures absorption in the banded region below the dissociation continuum is followed by emission of the fluorescence spectrum. There is, however, an electronic state, to which radiationless transitions are nominally 'forbidden', which crosses the $^3\Pi$ state, and addition of a foreign gas (Ar at a pressure of, say, 30 mmHg) is found to weaken the emission bands above the predissociation limit. Indeed, atomic iodine — the product of the predissociation — may be detected (by the atomic absorption lines) under those conditions where the emission bands are quenched. An increase in apparent intensity of the appropriate absorption bands has also been observed in the presence of a foreign gas, and there is little doubt in this case that the effect is a result of the induced predissociation. The predissociation leads to the formation of two *ground* state atoms, and can occur in the other halogens as well as in I_2 at wavelengths possessing enough energy to break the bond. Thus, from a study of the photobromination of ethylene, the quantum yield for the primary step

$$Br_2 + h\nu \rightarrow 2Br(^2P_{\frac{3}{2}}) \qquad (3.12)$$

has been shown to be unity for photolysis at wavelengths extending up to the energy of the dissociation limit (45·5 kcal (190kJ) mol^{-1} ≡ λ = 628·4 nm). In fact, the primary quantum yield is still unity at λ = 680 nm, where the energy (42·0 kcal (176 kJ) mol^{-1}) is insufficient to break the Br$_2$ bond. At this wavelength the extinction coefficient is temperature-dependent; the absorption appears to occur from levels with $v'' > 0$, and the photolysis affords another example of the contribution made to the energy required for dissociation by vibrational energy in the ground electronic state (cf. the example of NO$_2$ photolysis, pp. 50–52).

The S$_2$ molecule is interesting in that it has both a 'normal' predissociation (which has already been mentioned in Section 3.3) and an induced predissociation. Figure 3.3 shows that there is a $^1\Sigma_u^-$ state (dotted line) which crosses the $^3\Sigma_u^-$ state at $v' \sim 11$ (in $^3\Sigma_u^-$). Radiationless transitions to the $^1\Sigma_u^-$ state are forbidden by the $\Delta S = 0$ rule, but may be induced by molecular collisions. The crossing takes place at an energy rather above the S_2 dissociation limit for ground state atoms, and induced predissociation can occur: the predissociation is of subcase (a) (cf. Fig. 3.4). Diffuseness is, in fact, observed in the spectral region from 279·9 nm to 271·5 nm, and it is pressure-sensitive, the fine structure reappearing at sufficiently low pressures. The diffuseness at $\lambda < 261·5$ nm, corresponding to the $^3\Sigma_u^- \rightarrow \,^3\Pi_u$ normal predissociation, is *not* pressure-dependent.

A rather unusual feature of the induced predissociation in S$_2$ is the appearance of diffuseness at a wavelength (λ = 279·9 nm, $v' \sim 10$) rather below the crossing point ($v' \sim 11$) of the curves. The effect is looked upon as a result of quantum-mechanical 'tunnelling' through the potential barrier below the point of intersection; this seems to be the only well-established case of such tunnelling in predissociation.

One of the most dramatic examples of a predissociation induced by an external field, rather than by collision, is the quenching of iodine fluorescence in the presence of a magnetic field. The fluorescent emission in the visible spectrum is extinguished when a sufficiently strong magnetic field is applied. It has been shown that the selection rule $\Delta J = 0$ (3.8) no longer holds strictly in the presence of a magnetic field, and crossing can take place to one of the predissociating states which correlate with two ground state iodine atoms.

3.5 INTRAMOLECULAR ENERGY TRANSFER IN COMPLEX MOLECULES (1)

Radiationless transitions are favoured in complex molecules for the reasons suggested at the beginning of Section 3.3, and processes involving such intramolecular energy transfer are the most likely route to photodissociation. Detailed discussion of energy transfer is deferred until Chapters 4 and 5, since both intramolecular and intermolecular energy transfer are studied for the most part in terms of the emission phenomena described in Chapter 4. The processes of predissociation and induced predissociation have, however, been illustrated with examples involving very simple

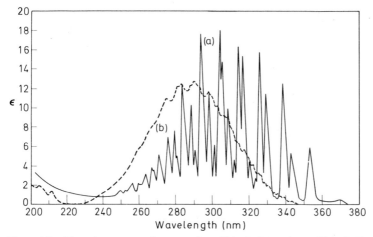

Figure 3.7. Absorption spectra of gaseous formaldehyde (a) and acetaldehyde (b), showing the loss of resolvable structure with increasing molecular complexity. (Derived from J. G. Calvert and J. N. Pitts, Jr., Photochemistry, *p. 368, Wiley, New York (1966))*

molecules, and we must see how far the photochemical dissociation of more complex species follows similar mechanisms.

Two main obstacles prevent description of the photochemistry of large molecules in such precise physical terms as those possible for simple molecules. First, the absorption spectrum of a complex species may not show sufficient resolvable structure either for an identification of state to be possible or for optical dissociation and predissociation to be recognised. The relative lack of spectroscopic structure is, of course, a result both of the greater complexity and closer spacing of vibrational and rotational levels, and of the increased number of electronic states. Figure 3.7 illustrates the loss of resolvable structure in going from formaldehyde to acetaldehyde. Secondly, a considerable number of fragmentation paths may exist for an excited polyatomic molecule. Although the relative importance of each path may depend on the wavelength of the dissociating radiation, several different sets of primary products are often produced simultaneously. In addition, reorganisation of the internal energy of an excited molecule can lead to rupture of a bond remote from the part of the molecule where the excitation was first localised. (In long-chain aldehydes, for example, cleavage can occur 'down-chain' from the carbonyl group which absorbs the energy: the cleavage may not even result directly from an electronically excited state, but rather from a high level of vibrational excitation in the ground electronic state). To illustrate the occurrence of several photodissociative steps, Table 3.2 gives the approximate quantum yields, where known, for five *primary* processes in the photolysis of n-butyraldehyde at different wavelengths (data taken from J. G. Calvert and J. N. Pitts, *Photochemistry*, pp. 372–3, Wiley, New York (1966)).

Table 3.2. Approximate quantum yields for primary steps in photolysis of n-butyraldehyde

Products	Quantum yield				
λ(nm) =	313·0	280·4	265·4	253·7	187·0
n-C_3H_7 + HCO	$\geqslant$0·35	$\geqslant$0·28	$\geqslant$0·28	0·31	0·34
C_3H_8 + CO	0·017	$\geqslant$0·11	$\geqslant$0·25	0·33	0·13
C_2H_4 + CH_3CHO	0·16	0·27	0·38	0·30	0·24
CH_3 + CH_2CH_2CHO	0·005	0·006	0·010	0·015	0·25
C_3H_6 + CH_2O	—	—	—	—	0·04

Although it may prove impossible to make direct spectroscopic assignments to the transition in an absorption, there are several ways in which the general type of the transition can be adduced. The maximum value of the measured extinction coefficient, ε_{max}, may be used to determine the degree of 'forbiddenness' in a transition. In Section 2.4 it was suggested that $\varepsilon_{max} \sim 10^5 \, cm^{-1} \, l \, mol^{-1}$ for a totally permitted transition (this value is dependent, of course, on the width of the absorption band, and is only an order-of-magnitude number). Thus the near-ultra-violet absorption bands ($\lambda \sim 240-340 \, nm$) of the aldehydes (cf. Fig. 3.7) have values of ε_{max} in the range 12-20 (all units are for molar decadic extinction coefficients), and are presumed to have their origins in a symmetry-forbidden transition (a singlet–singlet n–π* transition). Another absorption band at $\lambda \sim 400 \, nm$ is extremely weak ($\varepsilon_{max} \sim 10^{-3}$), and is probably due to a transition that is both spin- and symmetry-forbidden. Most organic molecules are singlets in their ground states, and very small values of ε_{max} may well suggest that the electronic states produced on absorption are triplets; evidence from emission processes can often confirm such a hypothesis. The oscillator strength (Section 3.5) for a perfectly allowed π–π* transition is thought to be of the order of unity. Table 3.3 gives a rough idea of the reduction in oscillator strength resulting from a contravention of certain selection rules; the factors are, of course, multiplicative if several selection rules are broken simultaneously.

Table 3.3. Oscillator strengths for various types of transition (cf. J. R. Platt, *J. opt. Soc. Am.* **43**, 252 (1953))

Type of transition	Oscillator strength	Specific system
Allowed	1	π–π*
Spin-forbidden ($\Delta S \neq 0$)	10^{-5}	second row elements
Overlap-forbidden (electronic charge changes position)	10^{-2}	n–π* in second row heteroatoms
Momentum-forbidden (large changes in linear or angular momentum)	$10^{-1}-10^{-3}$	condensed ring systems
Parity-forbidden (g ↮ u)	10^{-1}	condensed ring systems

In molecules for which both n–π* and π–π* transitions are possible the two possibilities may often be distinguished by the effect that polar and non-polar solvents have on the wavelengths of the absorption band. Only one n electron is available in an excited n, π* state, and polar solvents are more strongly hydrogen-bonded to the ground state than to the excited state. The energy of the ground state is therefore lowered more by the solvent interaction than is the excited state, the energy gap between the two levels is increased, and a shift of the absorption spectrum to shorter wave-lengths ('blue shift') is observed in polar solvents. A *red* shift is observed in some π–π^* transitions (e.g. in ketones) in polar solvents; the excited state is *more* polar than the ground state, and interaction with a polar solvent reduces the energy difference between the two states.

If it is possible to produce a single crystal, or an isotropic distribution of molecules in a rigid glass, of a molecule with a plane of symmetry, so that the orientation of the crystal plane to the molecular axes is known, then the interaction with polarised light can reveal the nature of the spectroscopic transition. For example, the π–π^* transition in pyridine is polarised in the plane of the molecule, while the n–π^* transition is polarised perpendicular to it. Kasha has given (in *Light and Life*, Johns Hopkins Press, Baltimore, 1961) a list of criteria for assignment of a transition as n–π^* rather than as π–π^*. These criteria are: (a) an n–π^* transition is absent in analogous hydrocarbons; (b) n–π^* bands disappear in acid media; (c) n–π^* bands show a 'blue' shift in polar solvents; (d) conjugative substituents also cause a blue shift; (e) n–π^* bands are found at relatively long wavelengths compared with π–π^* bands; (f) n–π^* bands are relatively weak compared with π–π^* bands; and (g) the transitions may show specific polarisation (e.g. the n–π^* transition in pyridine is polarised perpendicular to the plane of the ring).

The preceding discussion will have made it clear that potential energy curves (or, rather, sections through the many-dimensioned potential energy hypersurface) cannot usually be constructed for complex polyatomic molecules. Figure 3.8 shows an alternative diagram for a complex molecule such as naphthalene. The diagram is a modification of a form devised by Jablonski, and is frequently called a *Jablonski diagram*. It does not attempt to represent the molecular shapes and sizes, and the vibrational levels drawn for each state do not usually correspond to the actual v'', v' numberings and spacings. On the other hand, the energies of the vibrational ground states of each electronic level are shown correctly if the experimental evidence is available. As will be seen, the S_0, S_1, . . . , T_1, . . . notation is employed. Wavy lines on the diagram represent radiationless energy conversion: the vertical wavy lines within a particular electronic state indicate degradation of vibrational excitation (probably by a collisional, *inter*molecular process), while the horizontal wavy lines indicate intramolecular energy exchange. Formal distinction is drawn between electronic energy exchange

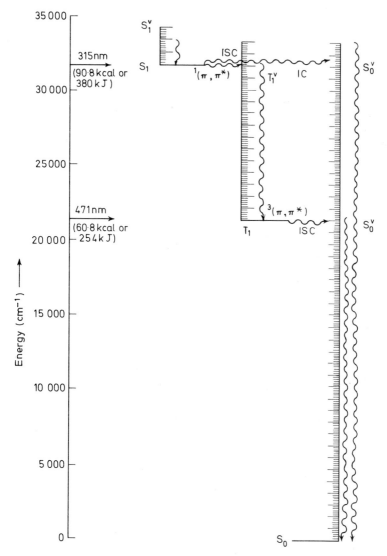

Figure 3.8. Method of representing energy levels and electronic states for a complex molecule such as naphthalene. This is a modified 'Jablonski diagram'

permitted by the $\Delta S = 0$ rule and that forbidden by it. The term *internal conversion* (IC) is applied to radiationless transitions between states of the same spin multiplicity, while *intersystem crossing* (ISC) refers to energy exchange between states belonging to different (spin) 'systems'. Both internal conversion and intersystem crossing are assumed to take place with no change in total electronic + vibrational energy, and the wavy lines are therefore horizontal (i.e. no translational or rotational energy is released in an intramolecular electronic energy exchange – see Chapter 4).

Jablonski diagrams are, in fact, more frequently used in connection with discussions of luminescence, but are introduced in this chapter to illustrate the similarity between predissociative fragmentation in simple molecules and the route to photodissociation in complex ones. The process of internal conversion from $S_1(v = 0)$ to $S_0^{v'}$ shown in Fig. 3.8 leads to a high degree of vibrational excitation in the ground electronic state. If sufficient energy is available, the molecule may then undergo spontaneous dissociation (or isomerisation, etc.). The condition of a molecule in our example (where $S_0^{v'}$ is populated) is identical to the condition of a molecule thermally 'activated' by collision. If enough energy to break a bond is stored somewhere in the molecule, then, so long as molecular collision does not *deactivate* the molecule, dissociation may occur when the energy has accumulated in the bond. The competition between reaction and deactivation, and the way in which energy accumulates in a bond, is treated by the several theories of unimolecular reaction (e.g. 'free-flow' or 'strictly harmonic'), and further discussion is out of place here. Suffice it to say that photochemical activation is an almost ideal technique for the production of virtually monoenergetic species with which the theories may be tested. Several examples are now known in which the rate of unimolecular decomposition shows an increase with decrease in wavelength of photolysis that is exactly predictable on a theoretical basis.

The essential differences between predissociation of a diatomic molecule and the mechanism we have just outlined for photodissociation of a complex entity should now become clear. If some state of a diatomic molecule is populated, by intramolecular energy transfer, with enough energy to dissociate, then it is likely to dissociate in the first vibration (i.e. in about 10^{-13} s); if it is populated at a level below the dissociation limit of this state, then (except under special circumstances) it will not dissociate. In a large molecule it is necessary to make three major modifications to this description. First, any one of a number of bonds may be broken, and, further, intramolecular elimination reactions can occur. Secondly, the time taken for the energy to flow to the bond(s) involved may be orders of magnitude greater than the 10^{-13} s taken for the diatomic molecule to dissociate, and the probability that the excited molecule may be deactivated before it can dissociate is increased accordingly. Thirdly, thermal energy already possessed in vibration of other bonds of the molecule can contribute to the total energy required for disruption of the reactive bond.

3.6 IMPORTANT PRIMARY DISSOCIATION PROCESSES

The chemical identity of the products of dissociation of diatomic species is immediately apparent. However, since the dissociation of a polyatomic molecule could yield many different products, an understanding of the photochemistry of large molecules requires that the nature of the primary products be known. Direct identification is sometimes possible; alternatively, the chemistry of the primary step may have to be inferred from the final products of reaction. It is not possible to describe here the problems associated with the elucidation of the primary dissociative mechanisms in particular molecules; for a more detailed discussion the reader is referred first to Chapter 5 of *Photochemistry* by Calvert and Pitts (Bibliography) and then to the original references cited. In this book a table is provided to summarise data for the photochemistry of inorganic species (Table 3.4), and a short description is given of the photochemistry of the major classes of organic compound.

The primary photochemical act in many organic compounds is followed so rapidly by decomposition, rearrangement or reaction of the intermediates first formed that the whole sequence of events is best considered as a primary process. We shall therefore point out the occurrence of these processes where appropriate, even though a general discussion of secondary thermal reactions is excluded.

Hydrocarbons

The alkanes absorb strongly in the 'vacuum' ultra-violet region: methane starts to absorb around 144 nm, and the higher alkanes absorb at progressively longer wavelengths (e.g. n-butane shows

Table 3.4. Primary dissociative processes in the photochemistry of some inorganic molecules. Data obtained from Chapter 3 of J. G. Calvert and J. N. Pitts, Jr., *Photochemistry*, Wiley, New York (1966), in which original references may be found

Species	Products[a]		Wavelength[b] (λ in nm)	Quantum yield[c]	Remarks
Hydrides					
H_2O	$H + OH(^2\Pi)$	(1)	<242	~ 1	$OH(^2\Sigma \rightarrow {}^2\Pi)$ fluorescence observed
	$H + OH(^2\Sigma)$	(2)	<135.6	—	
	$H_2 + O(^1D)$	(3)	123.6	$\phi_3/(\phi_1 + \phi_2) \sim 0.3$	
H_2O_2	$2OH(^2\Pi)$	(4)	253.7	0.85 ± 0.2	
	$OH(^2\Pi) + OH(^2\Sigma)$	(5)	<202.5	—	$OH(^2\Sigma \rightarrow {}^2\Pi)$ fluorescence observed
H_2S	$H + SH$	(6)	$200 - 255$	~ 1	
NH_3	$NH_2 + H$	(7)	<217	96% of reaction at $\lambda = 184.9$ nm	Total ϕ for NH_3 disappearance ~ 1 at $\lambda = 184.9$ nm and at low pressures. At $\lambda = 147$ nm, and at $p = 15$ mmHg, $\phi = 0.45 \pm 0.1$
	$NH(^3\Sigma^-) + 2H$	(8)	<155		
	$NH(^1\Pi) + H_2$	(9)	<129.5	14% of reaction at $\lambda = 123.6$ nm	
	$NH_3^+ + e$	(10)	<123		
HN_3	$N_2 + NH$	(11)	~ 200		
$NH_2.NH_2$	$NH_2 + NH_2$	(12)	199	~ 1	Possible alternative split to $H + NHNH_2$
HI	$H + I$	(13)	<327	~ 1	Fraction of I in $^2P_{\frac{1}{2}}$ and $^2P_{\frac{3}{2}}$ states at various wavelengths not known. Similar processes for HBr, HCl

Species	Products[a]		Wavelength[b] (λ in nm)	Quantum yield[c]	Remarks
Oxides					
O_2	$O(^3P)+O(^3P)$	(14)	~245·4		Via forbidden absorption to the $^3\Sigma_u^+$ state of O_2
	$O(^3P)+O(^1D)$	(15)	<175·9	~1	Via allowed absorption to $^3\Sigma_u^-$ in Schumann–Runge continuum
	$O(^3P)+O(^1S)$	(16)	<134·2		Suggested from appearance of spectrum. $2O(^1S)$ can be formed at $\lambda<92\cdot3$ nm
O_3	$O(^1D)+O_2(^1\Delta_g)$	(17)	<310	—	
	$O(^1D)+O_2(^1\Sigma_g^+)$	(18)	<266	~1	
	$O(^3P)+O_2(^3\Sigma_g^-)$	(19)	~600	~1	
SO_2	$SO+O$	(20)	<218	—	
SO_3	$SO_2+O(^3P)$	(21)	<344		At $\lambda<224$ nm, $O(^1D)$ can be formed
	$SO+O_2(^3\Sigma_g^-)$	(22)	<300		SO observed in flash photolysis but could be product of secondary reaction
N_2O	$N_2(^1\Sigma_g^+)+O(^1D)$	(23)	~180	~1	
	$N+NO$	(24)		12% at $\lambda = 123\cdot6$ nm	
NO	$N(^4S)+O(^3P)$	(25)	183·2		
	$N(^2D)+O(^3P)$	(26)	123 to 140		
	$N(^2P)+O(^3P)$	(27)	<123		
	$NO^+ + e^-$	(28)	<134·3		

(continued overleaf)

Table 3.4 (*continued*)

Species	Products[a]		Wavelength[b] (λ in nm)	Quantum yield[c]	Remarks
NO_2	$NO + O(^3P)$	(29)	<400	see discussion in Section 3.3 for variation of Φ with λ	
	$NO + O(^1D)$	(30)	228·8		
NOCl	$NO + Cl$	(31)	<760	~1 at $\lambda = 253·7$	
CO_2	$CO(^1\Sigma^+) + O(^1D)$	(32)	<165	~1 for $\lambda = 123·6–150$ nm	
	$CO(^1\Sigma^+) + O(^1S)$	(33)	<127·3	$= 1·2 \pm 0·1$ at 123·6 nm	
	$CO(^3\Pi) + O(^3P)$	(34)	<107		
C_3O_2	$CO + C_2O$	(35)	~300	—	
Cl_2O	$ClO + Cl$	(36)	220 to 850	—	
ClO_2	$ClO + O(^3P)$	(37)	~375·3		
	$ClO + O(^1D)$	(38)	—		
Halogens					
I_2	$I(^2P_{\frac{3}{2}}) + I(^2P_{\frac{1}{2}})$	(39)	<499	~1	Optical dissociation
	$2I(^3P_{\frac{3}{2}})$	(40)	<803·7		Induced predissociation
					Similar processes for other halogens

absorption at $\lambda < 166$ nm). The maximum decadic molar extinction coefficients are about 10^4, and the absorption is thought to be due to an allowed σ–σ^* transition. In the wavelength region 129·5 nm– 147 nm molecular elimination of hydrogen is the most important photodissociative process.

$$RCH_2R' + h\nu \rightarrow RCR' + H_2 \qquad (3.13)$$

although bond ruptures at almost any point are minor processes and can lead to the formation of hydrogen atoms and a variety of free radicals. At $\lambda = 123\cdot6$ nm process (3.13) is nearly six times more frequent than the radical fission (3.14):

$$CH_4 + h\nu \rightarrow CH_3 + H \qquad (3.14)$$

Photoionisation occurs at shorter wavelengths (for CH_4 with $\lambda < 96\cdot7$ nm).

The lowest energy singlet–singlet $(\pi$–$\pi^*)$ absorption band is found at longer wavelengths in unsaturated hydrocarbons than in the paraffins. For ethylene and substituted ethylenes the absorption maximum lies around 180 nm, while conjugation shifts the spectrum even further towards the visible: the compound $CH_3(CH=CH)_{10}$ CH_3 has an absorption maximum at $\lambda \sim 476$ nm. Isomerisation (Chapter 6) is a frequent fate of the excited state formed on absorption, but fragmentation is also observed. For example, in ethylene the processes

$$CH_2=CH_2 + h\nu \rightarrow H_2 + H_2C=C:(\rightarrow HC\equiv CH) \qquad (3.15)$$

$$\rightarrow 2H + H_2C=C: \qquad (3.16)$$

$$\rightarrow H_2 + HC\equiv CH \qquad (3.17)$$

$$\rightarrow 2H + HC\equiv CH \qquad (3.18)$$

are thought to take place over the wavelength range 123·6 nm– 184·9 nm. Acetylene is formed in (3.15), via $H_2C=C:$, about 1·5 times more rapidly than in the direct process (3.17) at $\lambda = 147$ nm. At the same wavelength the radical (3.16, 3.18) and molecular (3.15, 3.17) mechanisms occur with almost equal efficiency; fission to $CH_2CH. + H.$ is much less frequent.

Fragmentation of the polyenes is important in low-pressure gas-phase photolysis, but is suppressed in the presence of added gas. It would appear that bond rupture occurs in a vibrationally excited ground state molecule which is formed by intersystem crossing from the excited singlet state.

Simple aromatic hydrocarbons possess a moderately strong absorption in the near ultra-violet: for benzene the absorption is of the symmetry-forbidden $^1B_{2u} \leftarrow \,^1A_{1g}$ system and has $\varepsilon_{max} \sim 160$ at $\lambda_{max} = 256$ nm. An intense allowed absorption band is observed at shorter wavelengths (in benzene the transition of the $^1E_{2u} \leftarrow \,^1A_{1g}$ system lies at $\lambda \sim 180$ nm). Absorption in the longer-wavelength band leads mainly to emission of radiation (Chapter 4) or to reaction of the excited species (Chapter 6), and quantum yields for photodissociation are very small. However, in the gas phase at $\lambda = 184 \cdot 9$ nm the quantum yield for benzene disappearance is near unity, and polymer, carbon and traces of volatile products are formed.

Alkyl halides

The alkyl halides possess a first absorption band in the wavelength region 200 nm–300 nm, which is due to promotion of a non-bonding p electron on the halogen atom to an anti-bonding σ^* orbital. The transition is partially forbidden; it becomes more allowed, and shifts to longer wavelengths, with increased atomic weight of the halogen, and with increased halogenation. Table 3.5 shows the wavelength of maximum absorption, λ_{max}, and ε_{max}, for bromo- and iodo-methanes.

Table 3.5. λ_{max}(nm) and ε_{max}(cm^{-1} l mol^{-1}) for bromo- and iodo-methanes. Data from J. G. Calvert and J. N. Pitts, Jr., *Photochemistry*, pp. 522–523, Wiley, New York (1966).

| Species | Halogen | | | |
| | Br | | I | |
	λ_{max}	ε_{max}	λ_{max}	ε_{max}
CH_3X	203·0	264	257·6	380
CH_2X_2	220·0	1100	290·0	1320
CHX_3	224·0	2130	349·0	2170

Photolysis of the monohaloalkanes by radiation absorbed in this band leads to formation of a free radical and a halogen atom

$$RX + h\nu \rightarrow R^* + X \qquad (3.19)$$

with a quantum yield of unity. R^* represents a 'hot' alkyl radical whose excitation may be in all or some of electronic, vibrational or translational modes; if X is heavy, much of the excess energy of hv over the R–X dissociation energy is taken up by R*, and enhanced reactivity of such radicals has been observed. There is some evidence that carbon monohalides, found in the photolysis of the compounds CXYZBr (where X, Y, Z = H, F, Cl or Br) are produced from vibrationally 'hot' radicals, e.g.

$$CHClBr_2 + hv \rightarrow CHClBr^* + Br \qquad (3.20)$$

$$CHClBr^* \rightarrow CCl + HBr \qquad (3.21a)$$

$$\rightarrow CBr + HCl \qquad (3.21b)$$

In general, the photolysis of hetero-polyhalomethanes leads to rupture of the weakest C—halogen bond.

The alkyl iodides possess a second strong absorption band ($\varepsilon_{max} \sim 7700$) at $\lambda \sim 190$ nm which is due to an s ← p promotion of the non-bonding electron on the iodine. Photolysis in this region leads to molecular elimination of hydrogen iodide, e.g.

$$C_2H_5I + hv \rightarrow C_2H_4 + HI \qquad (3.22)$$

Electronically excited I_2 may be produced by molecular elimination in the vacuum ultra-violet photolysis of CH_2I_2:

$$CH_2I_2 + hv \rightarrow CH_2 + I_2\dagger \qquad (3.23)$$

Nitrites, nitrates and nitroso- and nitro-compounds

Alkyl nitrites show a structured absorption region between 300 nm and 400 nm with $\varepsilon_{max} \sim 60$, and a stronger continuous absorption at $\lambda \sim 220$ nm ($\varepsilon_{max} \sim 1500$). The important primary step appears to be

$$RONO \rightarrow RO^{(*)} + NO \qquad (3.24)$$

The quantum yield is unity (for *tert*-butyl nitrite) even in the structured absorption region. Alkyl nitroso-compounds are often unstable, but the photochemistry of substituted nitroso-compounds (e.g. CF_3NO) has been studied. A long-wavelength absorption region occurs between about 500 nm and 750 nm, and photolysis by light of this wavelength proceeds dominantly via

$$R–N=O \rightarrow R + NO \qquad (3.25)$$

There has been little detailed study of the alkyl nitrates; the process analogous to (3.24) occurs for methyl and ethyl nitrates:

$$CH_3ONO_2 + h\nu \rightarrow CH_3O^* + NO_2 \qquad (3.26)$$

Two minor processes have been established for ethyl nitrate:

$$C_2H_5ONO_2 + h\nu \rightarrow CH_3CHO + HONO \qquad (3.27)$$
$$\rightarrow C_2H_5ONO + O \qquad (3.28)$$

The nitroalkanes photodissociate mainly to a free radical and nitrogen dioxide,

$$CH_3NO_2 + h\nu \rightarrow CH_3 + NO_2 \qquad (3.29)$$

although in nitroalkanes with a hydrogen attached to the β-carbon molecular elimination of nitrous acid may also occur:

$$CH_3CH_2NO_2 + h\nu \rightarrow CH_3CH_2 + NO_2 \qquad (3.30a)$$
$$\rightarrow CH_2{=}CH_2 + HONO \qquad (3.30b)$$

Fission of the O—NO bond is more important than processes such as (3.29) or (3.30) in the photolysis of aromatic nitrocompounds:

$$C_6H_5NO_2 + h\nu \rightarrow C_6H_5N{=}O + O \qquad (3.31)$$

Oxygen-containing compounds: ethers, peroxides, alcohols

The ethers and alcohols show absorption of a partially forbidden ($\varepsilon_{max} = 10^2$ to 10^3) n–σ* transition at $\lambda < 200$ nm. In both classes of compound radical and molecular fragmentation may occur, and it is not usually possible to determine the relative importance of the two processes. The two sets of paths are illustrated for diethyl ether and for ethanol:

$$C_2H_5OC_2H_5 + h\nu \rightarrow C_2H_5^* + C_2H_5O^*$$
$$\longrightarrow CH_3 + CH_2O \qquad (3.32)$$
$$\longrightarrow H + C_2H_4$$
$$\rightarrow C_2H_5OH + C_2H_4 \qquad (3.33)$$
$$\rightarrow CH_3CHO + C_2H_6 \qquad (3.34)$$
$$C_2H_5OH + h\nu \rightarrow C_2H_5^* + OH$$
$$\longrightarrow C_2H_4 + H \qquad (3.35a)$$

$$\rightarrow C_2H_5O^* + H$$
$$ \longrightarrow CH_3 + CH_2O \qquad (3.35b)$$
$$\rightarrow C_2H_4 + H_2O \qquad (3.36)$$
$$\rightarrow CH_3CHO + H_2 \qquad (3.37)$$
$$\rightarrow CH_4 + CH_2O \qquad (3.38)$$

In methanol, it has been shown that the molecular elimination reaction

$$CH_3OH \rightarrow HCHO + H_2 \qquad (3.39)$$

is the dominant process for photolysis at wavelengths between 180 nm and 200 nm, and that formation of $CH_3O + H$ occurs with lower efficiency.

Alkyl peroxides possess a wide absorption band which has its onset around $\lambda = 350$ nm. At $\lambda > 230$ nm photodissociation yields alkoxy radicals which carry with them large excess energies (around 60 kcal (251 kJ) mol^{-1} at $\lambda = 300$ nm); extensive fragmentation of the radicals occurs:

$$ROOR' + h\nu \rightarrow RO^* + R'O^* \qquad (3.40)$$

Collisional deactivation of the excited radicals is observed at high gas pressures or in solution; at a fixed temperature and pressure radical fragmentation becomes relatively more important at shorter wavelengths (i.e. at higher excess energies). For photolysis at $\lambda < 230$ nm the primary process

$$ROOR' + h\nu \rightarrow RO_2^* + R' \qquad (3.41)$$

is also observed.

Amines, hydrazines, etc.

Aliphatic amines absorb appreciably at $\lambda < 250$ nm and the first absorption maximum at $\lambda \sim 220$ nm ($\varepsilon_{max} = 10^2 - 10^3$) is due to an n–σ* transition. The hydrazines absorb at rather longer wavelengths. Photodissociation in the primary aliphatic amines yields mainly H atoms and amino radicals:

$$RNH_2 + h\nu \rightarrow RNH + H \qquad (3.42)$$

Dissociation of N–C bonds is thought to be an additional process in the photolysis of secondary amines:

$$R_2NH + h\nu \rightarrow R + RNH \qquad (3.43)$$

and this type of cleavage becomes dominant in tertiary aliphatic amines. Aromatic amines tend to fluoresce, rather than dissociate, after absorption of radiation, but at elevated temperatures processes of types (3.42) and (3.43) are both observed

$$C_6H_5NH_2 + h\nu \rightarrow C_6H_5NH + H \qquad (3.44)$$

$$\rightarrow C_6H_5 + NH_2 \qquad (3.45)$$

Substituted hydrazines undergo photolytic cleavage of the N–N bond

$$C_6H_5NHNHC_6H_5 + h\nu_{\sim 250\ nm} \rightarrow 2C_6H_5NH \qquad (3.46)$$

It appears that azines, on the other hand, dissociate mainly by a molecular mechanism

$$CH_3CH{=}N{-}N{=}CHCH_3 + h\nu \rightarrow CH_3CN + CH_3CH{=}NH \qquad (3.47)$$

over the wavelength range 260 nm–350 nm ($\varepsilon_{max} \sim 86$ at $\lambda \sim 280$ nm); at shorter wavelengths (238 nm) radical fragmentation assumes nearly equal importance

$$CH_3CH{=}N{-}N{=}CHCH_3 + h\nu \rightarrow CH_3 + CH{=}N{-}N{=}CHCH_3 \qquad (3.48)$$

Azo-compounds, diazo-compounds and azides

Azoalkanes absorb weakly in the near ultra-violet (e.g. $\varepsilon_{max} \sim 4$ at $\lambda \sim 330$ nm for azomethane), and the transition appears to be an n–π^* type involving non-bonding nitrogen electrons and the anti-bonding π^* orbital of the double bond. A stronger ($\varepsilon \sim 10^3$) absorption is seen at $\lambda \sim 200$ nm. Both absorption systems are shifted to longer wavelengths in the diazoalkanes, and the first absorption may occur in the visible region (e.g. for diazomethane, $\varepsilon_{max} \sim 3$ at $\lambda \sim 400$ nm). The primary process in the photolysis of azoalkanes yields alkyl radicals and molecular nitrogen with a quantum yield near unity at low pressures. Decomposition appears to proceed via a vibrationally excited molecule, since at high pressures the quantum yield drops markedly:

$$RN=NR' + hv \rightarrow (RN=NR')^*[\rightarrow R + N_2R'] \rightarrow 2R + N \quad (3.49)$$

If the radical N_2R' is an intermediate, its lifetime is so short that it enters into no chemical reactions. Aromatic azo-compounds, and both aliphatic and aromatic azo-compounds in solution, undergo isomerisation rather than dissociation.

Aliphatic diazo-compounds undergo photolytic rupture of the C—N bond,

$$R_2C=N_2 + hv \rightarrow R_2C^* + N_2 \quad (3.50)$$

and the 'hot' substituted methylene may isomerise or decompose. In the case of diazomethane itself, methylene is the product. Both triplet, ground state, and singlet, excited products may be formed, photolysis at shorter wavelengths apparently favouring the singlet product. It is suggested that intersystem crossing to a triplet in excited diazomethane can compete with decomposition, and it is this triplet which yields triplet methylene:

$$CH_2N_2 + hv \rightarrow CH_2N_2\dagger(S_1) \rightarrow CH_2\dagger(\text{singlet}) + N_2$$
$$CH_2N_2\dagger\dagger(T_1) \rightarrow CH_2(\text{triplet}) + N_2 \quad (3.51)$$

However, it is not entirely clear whether triplet methylene is not, in fact, formed mainly by deactivation of a singlet CH_2 first formed, rather than via triplet diazomethane. The cyclic diazomethane, diazirine, yields predominantly methylene on photolysis at $\lambda \sim 313$ nm:

$$H_2C\underset{N}{\overset{N}{\diagdown\!\!\mid\mid}} + h\nu \longrightarrow CH_2 + N_2 \quad (3.52)$$

Diazonium salts yield a carbonium ion and molecular nitrogen on photolysis in highly polar solvents (e.g. water); free radical and molecular nitrogen may be formed in less polar solvents by electron transfer in the primary step.

Fission of an N—N bond to yield N_2 and a nitrene appears to be the major dissociative step in the photolysis of azides:

$$R-N=N^+=N^- + hv \rightarrow R-N: + N_2 \quad (3.53)$$

The nitrene may be excited (electronically or vibrationally), and undergo isomerization to an imine:

$$\text{n-}C_3H_7N{=}N_2 + h\nu \rightarrow \text{n-}C_3H_7N: + N_2$$
$$\downarrow \rightarrow \text{n-}C_2H_5CH{=}NH \tag{3.54}$$

In some cases 1, 5-abstraction of H may occur, followed by ring closure to give a pyrrolidine.

Compounds containing the carbonyl group

Photochemical data are probably more extensive for compounds containing the carbonyl, CO, group than for any other class of organic compound. In the interest of brevity, we shall confine our remarks here mainly to the photochemistry of aldehydes and ketones, since, broadly speaking, the acids, acid anhydrides, esters, and even amides undergo analogous photodissociative reactions.

In each class of carbonyl-containing compound, the first absorption is of the forbidden $n-\pi^*$ transition. In aliphatic aldehydes the maximum of this band lies near 290 nm (cf. Fig. 3.7), while in ketones it is displaced to slightly shorter wavelengths (~ 280 nm); aromatic substitution shifts the absorption to longer wavelengths (e.g. $\lambda_{max} \sim 340$ nm in benzophenone). The absorption lies at considerably shorter wavelengths in acids, anhydrides and esters (<250 nm) and in amides (<260 nm). Allowed $\pi-\pi^*$ and $n-\sigma^*$ transitions give rise to intense absorption bands at short wavelengths (e.g. around 180 nm and 160 nm for aldehydes).

R. G. W. Norrish and his co-workers were the first to make a systematic study of the photochemistry of aldehydes and ketones, and it is for this reason that certain possible reaction paths are commonly known as *Norrish Type I*, *Norrish Type II*, etc. (the general formulation was first given by Norrish and Bamford in *Nature*, **140**, 195 (1937)).

The important *dissociative* primary photochemical processes in ketones can be represented as

$$RCOR' + h\nu \rightarrow R + (COR')^*$$
$$\downarrow \rightarrow R' + CO \tag{3.55a}$$

$$\rightarrow (RCO)^* + R'$$
$$\downarrow \rightarrow R + CO \tag{3.55b}$$

$$R_2CHCR_2CR_2COR' \rightarrow R_2C{=}CR_2 + CR_2{=}C(OH)R'$$

$$\text{(3.56)}$$

$$\longrightarrow CR'R_2CHO$$

The radical-forming processes (3.55) are those known as Norrish Type I, while the intramolecular fission of the C–C bond α–β to the carbonyl group is the Type II process. A 'Type III' process has been postulated to account for the intramolecular formation of olefinic products in the photolysis of compounds where fission of the α–β bond cannot yield them. For example, the photolysis at $\lambda = 253.7$ nm of methyl isopropyl ketone yields propylene and acetaldehyde as major products

$$CH_3COCH(CH_3)_2 + h\nu \rightarrow CH_3CHO + C_3H_6 \qquad \text{(3.57)}$$

(at $\lambda = 313$ nm Type I fission is much more important).

The relative importance of the two paths (3.55a) and (3.55b) in Type I photolysis seems to depend on the wavelength of photolysis. Thus in the photolysis of methyl ethyl ketone,

$$\overset{a}{CH_3COC_2H_5 + h\nu \rightarrow CH_3CO + C_2H_5} \qquad \text{(3.58a)}$$

$$\overset{b}{\rightarrow CH_3 + COC_2H_5} \qquad \text{(3.58b)}$$

the relative contribution of paths a and b vary from 40:1 at $\lambda = 313$ nm to 2.4:1 at $\lambda = 253.7$ nm.

Some thermal equilibration of hot acyl radicals may occur, but almost immediate decomposition to alkyl radical and carbon monoxide becomes increasingly likely as the wavelength of photolysis is decreased.

Type II photolysis in ketones has been shown fairly conclusively to proceed via a cyclic six-membered intermediate. For example, in methyl propyl ketone the process may be represented by Eq. (3.59). The *enol* thus formed then tautomerises to acetone.

$$\text{(3.59)}$$

Notwithstanding the considerable amount of effort expended, it is still not entirely clear whether Type I or Type II processes arise specifically from triplet or singlet excited states of the ketone. There is good evidence to indicate that a triplet excited state is the precursor of the Type II reaction for photolysis of many aldehydes and ketones at $\lambda = 313$ nm. On the other hand, participation of singlet states, especially at shorter wavelengths, is not entirely excluded.

Aldehydes photodissociate according to a 'Type I', and where appropriate a 'Type II', mechanism. An intramolecular elimination reaction, not observed with ketones, also occurs:

$$RCHO + h\nu \rightarrow RH + CO \qquad (3.60)$$

For the simple aliphatic aldehydes, at a fixed wavelength the quantum yield of reaction (3.60) bears a constant ratio to the quantum yield for 'Type I' fission. The ratio is, however, wavelength-dependent, being less than 0·05 at $\lambda = 313$ nm, 0·34 at $\lambda = 280·4$ nm, 0·95 at $\lambda = 265·4$ nm and 1·22 at $\lambda = 253·7$ nm.

Cyclic ketones decompose under the influence of light in a manner which suggests that a biradical is formed in the primary step. Cyclopentanone, for example, is photodissociated in the gas phase in three ways:

$$\text{C}=\text{O} + h\nu \longrightarrow 2\text{C}_2\text{H}_4 + \text{CO} \qquad (3.61)$$

$$\longrightarrow \quad + \text{CO} \qquad (3.62)$$

$$\longrightarrow \quad \text{CH}_2=\text{CHCH}_2\text{CH}_2\text{CHO} \qquad (3.63)$$

The primary products may certainly be interpreted in terms of initial formation of $\cdot \text{CH}_2\text{CH}_2\text{CH}_2\text{CH}_2\text{CO} \cdot$. It appears that ethylene is formed directly from $\cdot \text{CH}_2\text{CH}_2\text{CH}_2\text{CH}_2 \cdot$ (derived from the primary biradical by loss of CO), rather than from decomposition of 'hot' cyclobutane after ring closure. There is, however, considerable evidence that vibrationally excited molecules are of importance in the gas-phase photolysis.

3.7 PHOTOCHEMISTRY IN SOLUTION

Much of what has been said about photochemical processes so far has referred specifically to gas-phase photochemistry. This seems to be an appropriate point at which to indicate the differences between processes in the gas-phase and those in condensed phases (particularly in solution).

Both the act of absorption and the subsequent fate of excitation may be profoundly affected by going from the gaseous to the liquid state. We shall deal with the spectroscopy and the primary processes in turn.

The influence of polar and non-polar solvents on different types of electronic transition has already been mentioned in Section 3.5. Molecular energies may be reduced by solvation, and the extent of this reduction depends both on the nature of the electronic state and on the nature of the liquid phase or solvent. Absorption intensities, especially of nominally forbidden bands, may be affected by perturbations brought about by the proximity of solvent molecules. Collision broadening, and the absence of well-defined rotational energy levels in the liquid phase, may make it difficult to demonstrate the existence of a predissociation, even of a simple molecule, by other than circumstantial evidence. Such evidence sometimes suggests that the wavelength of the onset of predissociation may be different in the gaseous and liquid states. Thus in going from gas phase to solution, the wavelength at which a photochemical process occurs, and the importance of that process relative to other concurrent processes, may be changed.

Several changes are also apparent in the fates of the electronically excited species formed on absorption. First, the large rate of molecular collision in condensed phases makes the probability of physical quenching of electronically (or vibrationally) excited species much higher than in the gas phase, and primary quantum efficiencies may be correspondingly small in solution. Secondly, *chemical* quenching of excited species may occur: that is, the excited molecule may be removed by chemical reaction. In solution the high concentration of solvent molecules may make this process competitive with fragmentation, or, indeed, any other loss process. (The reactions of excited species are discussed in Chapter 6.) Thirdly, the fragments formed in a dissociative process may themselves react with solvent every time they appear, and the *overall* step may then be regarded for practical purposes as the primary one. This kind of process is particularly important for free radicals produced in a hydrogen-containing solvent: hydrogenation of the radical effectively prevents the *free* radical appearing (although, of course, a solvent radical is usually created in the reaction). Thus the photolysis of di-isopropyl ketone in solution yields carbon monoxide, propane and butyraldehyde as the main Type I products. The overall efficiency of the process is low, for the reasons to be discussed in the next paragraph, although the quantum yield for the Type I decomposition increases with temperature, and reaches about 0·3 at 96°C. No C_6H_{14} or $C_3H_7CO.COC_3H_7$ products appear in the solution-phase photolysis, the C_3H_7 and C_3H_7CO fragments yielding products only by hydrogenation.

A very important factor in altering photochemical behaviour in the gas phase and in solution is the occurrence of *cage effects* in the latter phases. In the liquid phase, collisions between species are not uniformly distributed in time, but occur in sets, or 'encounters'. The colliding partners are enclosed in a solvent 'cage' which tends to prevent their separation by diffusion, and they may make several mutual collisions before leaving each other's sphere of influence. The rates of chemical reactions may be affected considerably by the occurrence of many collisions in an encounter. In particular, radical recombination may occur at every encounter since excess vibrational energy in a newly formed bond can be removed by collision with solvent molecules. 'Primary' recombination of radicals, formed by dissociation of an electronically excited molecule, is therefore of frequent occurrence. It occurs within 10^{-11} s of the generation of the radicals, before they have separated by a molecular diameter. ('Secondary' recombination of radicals takes place within 10^{-9} s of radical formation, but does not *necessarily* involve the identical pair of radicals formed by dissociation of one molecule.) This so-called *geminate recombination* of radicals within a solvent cage can greatly reduce the effective quantum yield for the dissociation process, since the reactant may be regenerated by primary radical recombination much more often than the radicals can escape from the solvent cage. The importance of the effect is related to the kinetic energy of the fragments formed on photolysis, as well as to the viscosity of the solvent. If the fragments have sufficient energy, they may force their way out of the cage, and the primary quantum yield of some solution-phase photolyses increases as the wavelength of photolysing radiation is decreased. Figure 3.9 shows the variation of apparent quantum yield for I atom formation in the photolysis of I_2 in hexane (25°C) at various wavelengths. Over the wavelength range 404·7 nm–735 nm the 'true' (gas-phase) quantum yield would be expected to be unity (the induced predissociation would be efficient at high pressures, and the bond dissociation energy of I_2 is equivalent to a wavelength of 803·7 nm), and the increased apparent efficiency of the primary process at short wavelengths is looked upon as a reflection of the increased kinetic energy with which the

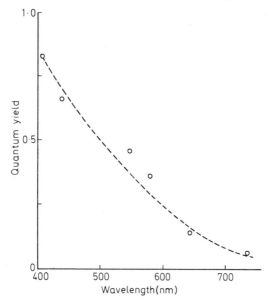

Figure 3.9. Apparent quantum yields for I atom formation on photolysis at different wavelengths of I_2 *in hexane solution at 25°C. (Data of L. F. Meadows and R. M. Noyes, J.* Am. chem. Soc. **82,** *1872 (1960))*

I atoms are born. The effect of solvent viscosity on the primary quantum yield is also consistent with this view. At $\lambda = 435 \cdot 8$ nm the quantum yield decreases from 0·66 in hexane (viscosity = 0·3 cP) to 0·036 when the viscosity is 180 cP.†

3.8   PHOTOCHEMISTRY OF IONIC SPECIES

A further distinctive feature of solution-phase photochemistry, especially where the solvent is water, is the importance of ionic species. 'Primary' photochemical reaction involving ions in solution

†For a more general discussion of cage effects and related diffusion phenomena, see I. D. Clark and R. P. Wayne in *Comprehensive chemical kinetics* (Ed. C. H. Bamford and C. F. H. Tipper), Vol. I, Elsevier, Amsterdam (1969).

is somewhat different from the photodissociations of neutral molecules, since the process often involves oxidation–reduction steps.

The colours of transition metal ions are in general the result of symmetry-forbidden optical transitions of electrons on the metal atom; the forbiddenness is revealed by relatively small oscillator strengths of around 10^{-4} (e.g. the absorption which causes the blue colour of the free Cu^{2+} ion has $\varepsilon_{max} \sim 10$ at $\lambda = 810\,m\mu$). Many ions do, however, have an *intense* absorption ($\varepsilon_{max} \sim 10^4$) in the ultra-violet region (usually, but not always, in the wavelength range 200 nm–250 nm). It is fairly clear that these absorptions derive from *charge-transfer* transitions: that is, from transitions in which an electron is transferred either from one ion to another, or from an ion to the solvent. The dark colour of the complex halides of Cu^{2+}, as distinct from the light blue of isolated Cu^{2+}, is due to the tail of such a charge-transfer spectrum which lies mainly in the ultra-violet.

We shall see shortly that, in many cases, the charge transfer process involves solvation of an electron removed from an ion. However, some early work on the alkali metal halides will be mentioned first, as it gave rise to the suggestion that the spectra were, indeed, due to charge transfer. Colour centres develop in crystals of alkali halides exposed to ultra-violet radiation ($\lambda \sim 200\,nm$), and it was suggested that these colours were due to free metal atoms in the lattice. The quantum yield for colour centre production is highly dependent on temperature, although the absorption spectrum is virtually temperature-independent (e.g., in KBr, the quantum yield varies from effectively zero to unity over the temperature range from $-100°C$ to $+400°C$). Thus the primary absorption process is the same at all temperatures, but the production of colour centres depends on an interaction with crystal lattice vibrations within the short lifetime of the excited state. It was suggested that dissociation occurred from an essentially covalent non-bonding state produced by irradiation of the alkali halide

$$K^+ Br^- + h\nu \rightarrow (KBr)\dagger \rightarrow K + Br \tag{3.64}$$

The absorption spectra of solid alkali halides lie in or near the wavelength region 160 nm–220 nm, and, in the case of the bromides and iodides, possess two main peaks. The separation of these two peaks corresponds closely to the energy difference between the $^2P_{\frac{3}{2}}$ and $^2P_{\frac{1}{2}}$ states of the free halogen atoms, and is regarded as good evidence that a free atom is formed in reaction (3.64). (In the absorption spectra of rubidium salts two bands are observed which are separated by an energy corresponding to the energy difference between Rb^2S and $Rb\dagger^2P$.) Although this picture of colour-centre formation in crystals has been modified, the process (3.64) almost certainly represents the situation in the photolysis of *gaseous* alkali halides. The gas-phase absorptions lie at longer wavelengths than in the solid state (presumably since it is no longer necessary to provide the energy difference between a normal crystal and the one formed after charge transfer), but for bromides and iodides there are still two peaks separated by the energy difference of $J = \frac{1}{2}$ and $J = \frac{3}{2}$ states in the free halogen atom.

In aqueous solution similar spectra are observed: at least for the iodides, the separation between the absorption maxima corresponds to the I atom excitation energy. That charge transfer is involved is strongly suggested by the correlation between the wavelength of the absorption and energies of charge-transfer processes for the same ion. For example, the energy of the long-wavelength edge of the bands in bivalent transition metal ions follows closely the redox potential of the bivalent–trivalent ion system. Again, a direct relationship exists between the energy corresponding to the absorption peak and the free energy of the reaction

$$X^{n+} + H^+ \rightarrow X^{(n+1)+} + \tfrac{1}{2}H_2 \qquad (3.65)$$

in many cases. Figure 3.10 shows this relationship for several ions: the slope is near unity.

Ionising radiation can strip an electron from water (the process is *radiolysis*), and a comparison of the reducing species formed by radiolysis with that produced by the photolysis of aqueous solutions of ions has suggested that the species is, in fact, a hydrated electron. A transient absorption at around 700 nm is observed on flash photolysis (see Chapter 7) of aqueous ionic solutions, and it appears identical with the absorption seen on pulse radiolysis of pure water. The rates of reaction of the species produced in the two ways are also often identical. Further, optical and electron paramagnetic resonance spectra of UV-irradiated aqueous glasses containing ionic species, and of trapped electrons produced by ionising radiation, are identical. It appears, therefore, that hydration of the electron can make photodetachment of an electron energetically favourable at wavelengths much longer than those needed to cause photoionisation in the gas phase. Quantum yields

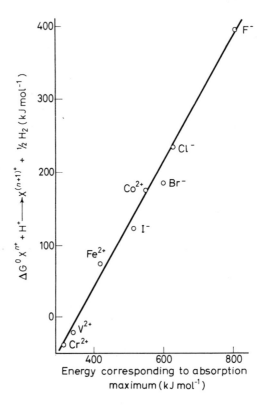

Figure 3.10. Graph to illustrate the direct relationship between the energy corresponding to absorption maximum of a charge-transfer spectrum and the free energy of the reaction

$$X^{n+} + H^+ \longrightarrow X^{(n+1)+} + \tfrac{1}{2}H_2$$

The nature of the ion from which the electron is transferred is indicated in the figure. (Data of R. J. Marcus, Science, **123**, *399 (1956))*

for the formation of hydrated electrons have been estimated, and may be relatively high. For photolysis of the halide ions (I^-, $\lambda = 253 \cdot 7$ nm; Br^-, Cl^-, $\lambda = 184 \cdot 9$ nm), for example, the quantum yields are probably $0 \cdot 3$–$0 \cdot 5$. The quantum yield for the process

$$Fe(CN)_6^{4-} + h\nu \rightarrow Fe(CN)_6^{3-} + e_{aq} \tag{3.66}$$

may even approach unity ($\lambda = 200$ nm–260 nm), although it is interesting that the quantum yield ($\lambda = 253 \cdot 7$ nm) for

$$Fe^{2+} + h\nu \rightarrow Fe^{3+} + e^- \tag{3.67}$$

is less than $0 \cdot 1$.

If the hydrated electron is involved, then the photochemical process may be written

$$X^{n+} + h\nu \xrightarrow{\text{H}_2\text{O}} X^{(n+1)+} + H_2O^- \tag{3.68}$$

The hydrated electron may then spontaneously dissociate,

$$H_2O^- \rightleftharpoons OH^- + H \tag{3.69}$$

or more particularly, in acid solution it may react according to

$$H_2O^- + H^+ \rightarrow H_2O + H \tag{3.70}$$

In either case atomic hydrogen is produced, and it can initiate secondary radical reactions. The photochemical decomposition of the I^- ion, for example, shows a dependence on pH which is consistent with a chain mechanism initiated by

$$I^- + h\nu \xrightarrow{\text{H}_2\text{O}} I + H_2O^- \rightarrow I + H + OH^- \tag{3.71}$$

and the low *overall* efficiency can be attributed to recombination in the cage of H and I atoms.

Charge transfer in cations can occur either to or from the solvent. Thus absorption bands for both types of charge transfer† are known in iron-containing solutions:

$$Fe^{2+}H_2O + h\nu \rightarrow Fe^{3+}H_2O^-, \quad \lambda_{max} \sim 285 \text{ nm} \tag{3.72}$$
$$Fe^{3+}H_2O + h\nu \rightarrow Fe^{2+}H_2O^+, \quad \lambda_{max} \sim 230 \text{ nm} \tag{3.73}$$

There is some evidence which suggests that the longest wavelength

---

†Although equations showing transfer of charge are *written* as if an electron is completely transferred, it must be understood that they may also represent partial electron transfer.

absorption usually corresponds to that direction of the oxidation–reduction reaction which proceeds most easily. The reactions of H_2O^+, analogous to (3.69) and (3.70), are suggested to be

$$H_2O^+ \rightleftharpoons OH + H^+ \tag{3.74}$$

and

$$H_2O^+ + OH^- \rightarrow OH + H_2O \tag{3.75}$$

In systems containing ion-pair complexes the energetics may be favourable for charge transfer from one partner (usually the cationic atom) to the other. The absorption maxima for the ion-pair charge-transfer spectra generally lie at longer wavelengths than the maxima for the uncomplexed cations (e.g. for the complex $Fe^{3+}CNS^-$, $\lambda_{max} \sim 460$ nm, compared with $\lambda_{max} \sim 230$ nm for charge transfer to Fe^{3+} from water). Typical processes involving Fe^{3+} complexes are:

$$Fe^{3+}CNS^- + h\nu \rightarrow [Fe^{2+}CNS] \rightarrow Fe^{2+} + CNS \tag{3.76}$$

$$Fe^{3+}Cl^- + h\nu \quad \rightarrow [Fe^{2+}Cl] \quad \rightarrow Fe^{2+} + Cl \tag{3.77}$$

$$Fe^{3+}OH^- + h\nu \quad \rightarrow [Fe^{2+}OH] \quad \rightarrow Fe^{2+} + OH \tag{3.78}$$

Atom production in (3.77) has been demonstrated by the photoinitiation of vinyl polymerisation by a radical mechanism in a ferric chloride system; it has also been shown that photochemical oxidations may be initiated by reaction (3.78) followed by hydrogen abstraction,

$$OH + RH \longrightarrow H_2O + R \tag{3.79}$$

Charge transfer occurs in two substances, uranyl oxalate and potassium ferrioxalate, which are frequently used as *chemical actinometers* (see Chapter 7). In uranyl oxalate charge transfer from uranyl to oxalate ions leads to decomposition of the oxalate; in potassium ferrioxalate the change of importance is the reduction of Fe^{3+} ions to Fe^{2+}.

BIBLIOGRAPHY

J. G. CALVERT and J. N. PITTS, JR., *Photochemistry:* Chapter 3, 'Interaction of light with simple molecules'; Chapter 5, 'Photochemistry of the polyatomic molecules'. Wiley, New York (1966).

A. G. GAYDON, *Dissociation energies and spectra of diatomic molecules:* Chapter IV, 'Photodissociation'; Chapter V, 'The Birge–Sponer extrapolation; Chapter VI, 'Predissociation'. Chapman and Hall, London (1953).

I. D. CLARK and R. P. WAYNE, 'The theory of elementary reactions in solution', in *Comprehensive Chemical Kinetics,* Vol. I (Ed. C. H. BAMFORD, and C. F. H. TIPPER), Elsevier, Amsterdam (1969).

L. E. ORGEL, 'Charge-transfer spectra and some related phenomena', *Q. Rev. chem. Soc.* **8,** 422 (1954).

4

Emission processes (1)

4.1 LUMINESCENCE

The emission of radiation from excited species is one of the several paths by which the excess energy may be lost (path vi, Fig. 1.2); the general phenomenon of light emission from electronically excited species is known as *luminescence*. In this chapter and the next we shall discuss luminescent processes. First, simple luminescent phenomena are considered, and then, in Chapter 5, *sensitised luminescence* is described: the latter process involves *inter*molecular energy transfer (path iv, Fig. 1.2) and electronic excitation is produced in a species other than the one which was initially excited. *Intra*molecular energy transfer, which populates a different *state* of the same *species*, is discussed in the present chapter.

Luminescent emission provides some of the most reliable information about the nature of primary photochemical processes. Competition exists between emission and other fates of excited species (quenching, reaction, decomposition, etc.), and the dependence of emission intensity on temperature, reactant concentrations, and so on, may yield valuable data about the nature and efficiencies of the various processes. In particular, quenching by bimolecular collisions, and unimolecular energy degradation by radiationless transitions, are almost always best studied in terms of their effect on the intensity of luminescence. As well as possessing this fundamental interest, luminescent phenomena are also of considerable importance in several commercial and scientific applications, and an example will be given in Chapter 8.

The various individual luminescent phenomena are named according to the mode of excitation of the energy-rich species. We are concerned primarily with excitation by absorption of radiation, and emission from species excited in this way is referred to as *fluorescence* or *phosphorescence*: the distinction between the two processes is discussed below. Emission following excitation by chemical reaction (of neutral or charged species) is known as *chemiluminescence*, and is described briefly in Section 4.7. Other means of providing electronic excitation, which will not be discussed further, are by heat (e.g. in NO_2 – *pyroluminescence*), by an electric field (e.g. in solid ZnS – *electroluminescence*), by electron impact in gases (e.g. in discharge lamps), by electron impact on solid phosphors (e.g. in television tubes – *cathodoluminescence*), by crushing crystals (e.g. uranyl nitrate – *triboluminescence*), and by rapid crystallisation from solution (e.g. strontium bromate – *crystalloluminescence*). Although we shall have occasion to refer to the luminescence of substances trapped in rigid glasses, we shall omit general discussion of the luminescence of solids. The emission of radiation from solids, especially inorganic compounds, is a complex phenomenon, but of the greatest importance (for example, the advent of colour television has provided a great stimulus to research on inorganic phosphors possessing emissions of specific colours and intensities). For an introduction to such luminescence, the reader is referred to Chapter 5 of *Luminescence in chemistry* (Bibliography).

The two emission processes in which the ultimate source of excitation is absorption of radiation – fluorescence and phosphorescence – were originally distinguished in terms of whether or not there was an observable 'afterglow'. That is, if emission of radiation continued after the exciting radiation was shut off, the emitting species was said to be phosphorescent, while if emission appeared to cease immediately, then the phenomenon was one of fluorescence. The essential problem is what is meant by 'immediately' in this context, since the observation of an afterglow will obviously depend not only on the actual rate of decay of the emission (see Section 4.2 for further discussion of emission lifetimes), but also on the techniques used to observe it. Various instruments were devised to observe 'short-lived' luminescence, and in the early 1930s a luminescence with a lifetime less than about 10^{-4} s was thought to be short-lived and, hence, fluorescent. In 1935 Jablonski interpreted phosphorescence as being emission from some long-lived metastable electronic state lying lower in energy than the state populated by absorption of radiation (cf. pp. 55–61, Chapter 3). Several workers (among them Lewis and Kasha, and Terenin) suggested that the long-lived metastable state was, in fact, a triplet state of the species, and, as we shall see in Section 4.4, there is now considerable experimental evidence to substantiate this hypothesis.

The long lifetime of the emission is a direct consequence of the 'forbidden' nature of a transition from an excited triplet to the ground state singlet; that electric dipole transitions occur at all where $\Delta S \neq 0$ is due to the inadequacy of S to describe a system in which there is spin–orbit coupling (cf. Section 2.7). Extension

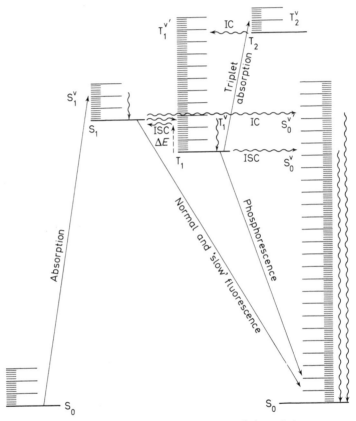

Figure 4.1. Jablonski diagram showing absorption, and the emission processes of fluorescence and phosphorescence

of this idea to other systems, not necessarily triplet–singlet, in which $\Delta S \neq 0$ leads to the useful definition of phosphorescence as a *radiative transition between states of different multiplicities*; fluorescence is then understood to be a radiative transition between states of the *same* multiplicity. Figure 4.1 is a Jablonski diagram showing the processes of fluorescence and phosphorescence. These definitions are used almost universally by organic photochemists, although they might be extended, to include in 'phosphorescence' emission processes involving a transition forbidden by *any* selection rule, rather than just the $\Delta S = 0$ rule. Since the distinctions between allowed and forbidden transitions are not sharp, the definitions lack some precision.

Absorption of radiation in a singlet–triplet transition is weak, since it is forbidden in the same way as the triplet–singlet phosphorescent emission. It follows that phosphorescence can only be excited inefficiently by direct absorption of radiation, and phosphorescence is much more usually the result of emission from a triplet populated by intersystem crossing from an excited singlet formed on absorption. In Fig. 4.1 absorption populates S_1^v; vibrational energy, at least in condensed phases, is rapidly degraded, and S_1^0 can then lose its energy by radiation, intersystem crossing (ISC) to T_1^v or internal conversion (IC) to S_0^v. It is, perhaps, surprising that ISC to T_1^v, which is spin-forbidden (radiationless transition selection rule), can compete effectively with spin-allowed fluorescence and IC to S_0^v; phosphorescence is, however, observed in many systems and, together with other evidence, this suggests that IC from $S_1 \rightsquigarrow S_0$ is relatively inefficient. A complete understanding of the photochemistry of a molecule really requires that the efficiencies (i.e. quantum yields) be known for all the processes occurring. Even if chemical reaction, decomposition and physical quenching of an excited species do not occur, it is still necessary to measure quantum yields for fluorescence (ϕ_f), phosphorescence (ϕ_p), intersystem crossing $T_1 \rightsquigarrow S_0$ (ϕ_{ISC}) and internal conversion $S_1 \rightsquigarrow S_0$ (ϕ_{IC}). With the restrictions on the processes occurring, it follows that

$$\phi_f + \phi_p + \phi_{ISC} + \phi_{IC} = 1 \qquad (4.1)$$

(although the relative magnitudes of the four quantum yields may be affected by the external environment). Unfortunately, these quantum yields have not been measured in many instances, and there is, therefore, often an uncertainty about the relative importance of the several primary processes. In Chapter 7 we shall see what techniques are available for the measurement of these quantum yields. To add to the problem, published efficiencies of the various processes cannot always be taken to be accurate. Trace impurities can lead either to an enhancement or to complete quenching of the emission phenomena. For example, the blue fluorescence of fluorene may be entirely dependent on traces of carbazole produced during manufacture, while the measured first-order rate constant for ISC ($T_1 \rightsquigarrow S_0$) in anthracene solutions has dropped from $1300 \, s^{-1}$ to essentially zero as solutions increasingly free of oxygen have been employed. The physical condition, which includes the phase, of a species can also markedly affect the rates of the several competing processes, and comparison of data obtained from experiments performed in different phases must be carried out only with the utmost caution, if at all. Apparent lifetime measurements are, of course, subject to the same problems as determinations of quantum yields, and should be treated with a similar reserve.

4.2   KINETICS AND QUANTUM EFFICIENCIES OF EMISSION PROCESSES

Considerable information about the efficiencies of radiative and radiationless processes can be obtained from a study of the kinetic dependence of emission intensity (or quantum yield) on concentrations of emitting and quenching species. In this section we shall consider first the application of stationary state methods to fluorescence (or phosphorescence) quenching, and then discuss the lifetimes of luminescent emission under non-stationary conditions.

Observable effects in the quenching of fluorescence are usually the result of competition between radiation and bimolecular collisional deactivation of *electronic* energy, since vibrational relaxation is normally so rapid, especially in condensed phases, that emission derives almost entirely from the ground vibrational level of the upper electronic state: this point is discussed further in the next section. The simplest excitation–deactivation scheme, which does not allow for intramolecular radiationless processes, is

$$\text{rate:}$$

$$X + h\nu_{abs} \rightarrow X^* \qquad \text{absorption} \qquad I_{abs} \qquad (4.2)$$

$$X^* + M \xrightarrow{k_q} X + M \qquad \text{quenching} \qquad k_q[X^*][M] \qquad (4.3)$$

$$X^* \rightarrow X + h\nu_{em} \qquad \text{emission} \qquad A[X^*] \qquad (4.4)$$

Solution of the steady state equations for $[X^*]$ (i.e. with

$d[X^*]/dt = 0$) leads to the result that

$$I_{\text{emitted}} = A[X^*] = \frac{AI_{\text{abs}}}{A + k_q[M]} \quad (4.5)$$

(note that the rate constant A is the Einstein coefficient for spontaneous emission). Equation (4.5) can be inverted to give the *Stern–Volmer* relation

$$\frac{1}{I_{\text{emitted}}} = \frac{1}{I_{\text{abs}}}\left(1 + \frac{k_q}{A}[M]\right) \quad (4.6)$$

Figure 4.2 shows a typical plot of $1/I_{\text{emitted}}$ against $[M]$ for the quenching of the fluorescence of an aqueous quinine sulphate

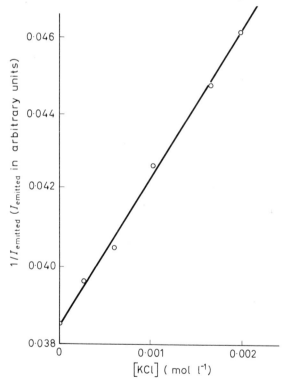

Figure 4.2. Stern–Volmer plot for the quenching of quinine sulphate fluorescence by the chloride ion. (Averaged points from practical course experiment, Physical Chemistry Laboratory, Oxford)

solution by the chloride ion. The values of the slope and intercept can be used to give a value of k_q/A even if $I_{emitted}$ is measured in arbitrary units and I_{abs} is not determined. Thus, if the Einstein A factor is known, or can be measured, the value of the quenching rate constant can be calculated. The A factor can be calculated from the B factor by use of the 'v^3' relationship derived in Section 2.3 (and B itself can, of course, be calculated from the measured integrated extinction coefficient for the absorption band, as described in Section 2.5). It is also possible, under suitable conditions, to measure A directly by observation of the decay of emission after suddenly extinguishing the illuminating beam. As we shall see later in this section, the fluorescence or phosphorescence lifetime may be shorter than the 'natural' radiative lifetime as a result of intermolecular and intramolecular non-radiative energy degradation, so that due care must be taken in the interpretation of emission decay measurements. (Very rarely, as mentioned in Section 4.6, the observed lifetime may be *longer* than the natural radiative lifetime.)

Rate constants for quenching can be compared with those predicted by the *collision theory* of chemical kinetics. According to this theory, a rate constant, k, is given by

$$k = \sigma_{coll}^2 \sqrt{\frac{8\pi \mathbf{k} T}{\mu}} \cdot e^{-E_a/RT} \qquad (4.7)$$

Table 4.1. Cross-sections for the quenching of NO_2 fluorescence by several gases. Data of G. H. Myers, D. M. Silver and F. Kaufman, *J. chem. Phys.* **44**, 718 (1966)

Quenching gas (M)	$10^{16}\pi\sigma_q^2$ (cm^2 molecule^{-1})	$10^{16}\pi\sigma_{coll}^2$	$\dfrac{\pi\sigma_q^2}{\pi\sigma_{coll}^2}$
He	1·6	34·2	0·04
Ar	3·8	44·0	0·08
N_2	5·0	48·7	0·10
O_2	6·0	46·2	0·13
H_2	2·5	38·3	0·06
NO	9·7	47·4	0·20
CH_4	7·9	50·2	0·16
N_2O	11·9	50·2	0·24
NO_2	13·5	55·3	0·24
CO_2	13·8	51·5	0·27
SF_6	25·7	72·5	0·35
CF_4	24·5	61·9	0·39
H_2O	28·3	59·3	0·49

Excitation wavelength = 435·8 nm. A taken to be $2·3 \times 10^4$ s^{-1}. (There is some reason, discussed in Section 4.6, to believe that this value is too small; a higher value might lead to correspondingly larger values of $\pi\sigma_q^2$.)

(σ_{coll} is half the sum of the gas-kinetic collision diameters of the reaction partners, and μ is their reduced mass; E_a is the activation energy for the reaction). E_a is expected to be near zero for collisional quenching, so one way of making the comparison is to calculate, from k_q, the *quenching cross-section* (which we will write as $\pi\sigma_q^2$) and compare it with $\pi\sigma_{coll}^2$. Table 4.1 shows some data obtained for the quenching of the fluorescence of NO_2 excited by light of $\lambda = 435 \cdot 8$ nm, together with values of the gas-kinetic cross-section. The ratio $\pi\sigma_q^2/\pi\sigma_{coll}^2$ corresponds to the familiar P factor of the collision theory (assuming that $E_a = 0$); the results suggest that quenching is a relatively efficient process for all quenching species, M, but that the efficiency increases with increasing complexity of M (note especially the effectiveness of water: polar molecules frequently seem to be highly efficient as physical deactivators). Even for M = He, only about 25 gas-kinetic collisions are needed, on average, to bring about quenching.

The relatively great rates of the quenching process may mean that in solution the rate is determined more by the rate of diffusion of quenching and emitting molecules than by the rate of collision. An *approximate* expression for the diffusion-limited rate constant, k_{diff}, is given by the Debye equation

$$k_{diff} \sim \frac{8RT}{3\eta} \times 10^3 \, \text{l mol}^{-1} \, \text{s}^{-1} \qquad (4.8)$$

where η is the viscosity of the solvent in newton-seconds per square metre and $R = 8 \cdot 3 \, \text{J K}^{-1} \, \text{mol}^{-1}$. For water at room temperature, $\eta \sim 10^{-3} \, \text{N s m}^{-2}$, so that $k_{diff} \sim 10^{10} \, \text{l mol}^{-1} \, \text{s}^{-1}$. The results for the quenching of quinine sulphate fluorescence (Fig. 4.2) give a value for $k_q/A \sim 100 \, \text{l mol}^{-1}$. For the transition involved, $A \sim 4 \cdot 3 \times 10^7 \, \text{s}^{-1}$, so that k_q approaches the diffusion-controlled limit. Note that, since η is temperature-dependent, k_q may increase with temperature so that there *appears* to be an activation energy for the process; however, the true E_a, in the sense of the energy needed for reaction once a collision occurs, can still be zero. Although the quenching rate approaches the diffusion-controlled limit, it is not necessarily true that every molecular *collision* leads to deactivation. The diffusive process limits the rate at which the excited species and the quencher come together, but prolongs each *encounter* so that several hundred collisions are possible before the two species diffuse apart (see also Section 3.7).

Quenching of luminescence by ground state atoms or molecules of the emitting species can occur, of course, and the process is known as *self-quenching*. If self-quenching is the only deactivation process, then the emitted intensity (but not the quantum yield) may be independent of concentration over some concentration range. This curious result is observed with nitrogen dioxide fluorescence, where the intensity of the orange-coloured emission excited by the blue mercury line (435·8 nm) can be seen not to vary with pressures between about 10^{-2} and 10 torr. The explanation is to be sought in terms of Eq. (4.5), but with the additional special condition that at low pressures and, hence, fractional absorption (see Eq. 2.16 in Section 2.4)

$$I_{abs} = I_0 \alpha [X] d \tag{4.9}$$

Substituting Eq. (4.9) into Eq. (4.5), and writing $[M] = [X]$ for self-quenching, we obtain the result

$$I_{emitted} = \frac{I_0 \alpha d [X]}{1 + \dfrac{k_q}{A}[X]} \tag{4.10}$$

If $(k_q/A)[X] \gg 1$, which it is at pressures at least down to 10^{-2} torr, then $I_{emitted}$ is independent of $[X]$.

The kinetics of the emission process has been developed in terms of excitation, emission and collisional deactivation steps. If intra-molecular energy-loss processes take place, then additional first-order terms must be added to the denominator of Eq. (4.5). Thus, if the rate constant which describes first-order energy degradation is k_1, the modified form of Eq. (4.5) is

$$I_{emitted} = \frac{A I_{abs}}{A + k_1 + k_q [M]} \tag{4.11}$$

In this case a Stern–Volmer plot ($1/I_{emitted}$ vs.$[M]$) will have an intercept of $(1 + k_1/A) I_{abs}^{-1}$ and slope $(k_q[M]/A) I_{abs}^{-1}$, and k_q/A cannot be determined unless $I_{emitted}$ and I_{abs} are measured. Only the *ratio* $I_{emitted}/I_{abs}$ is needed, so *absolute* values of intensity are not necessary, although arbitrary scales must be corrected to relative photon (rather than energy) fluxes if the wavelength distributions of exciting and emitted radiation are not identical. The ratio $I_{emitted}/I_{abs}$ is, of course, the quantum yield, ϕ_l, for the luminescent process (ϕ_f, fluorescence quantum yield; ϕ_p, phosphorescence quantum yield), and the intercept of a modified Stern–Volmer plot of $1/\phi_l$ vs. $[M]$ will indicate the relative ratio of radiative and radiationless loss processes. Hence, ideally it is possible to determine the rate of internal conversion or intersystem crossing in luminescent systems.

If a species exhibits both fluorescence and phosphorescence, and the phosphorescent state T_1 is excited by intersystem crossing from S_1, we derive the quantum yields from the following reaction scheme:

$$S_0 + h\nu \longrightarrow S_1 \qquad \text{excitation} \qquad (4.12)$$

$$S_1 + M \xrightarrow{k_1} S_0 + M \qquad \text{fluorescence quenching} \qquad (4.13)$$

$$S_1 \xrightarrow{A'} S_0 + h\nu' \qquad \text{fluorescence} \qquad (4.14)$$

$$S_1 \xrightarrow{k_2} T_1 \qquad \text{ISC} \qquad (4.15)$$

$$S_1 \xrightarrow{k_3} S_0 \qquad \text{IC} \qquad (4.16)$$

$$T_1 + M \xrightarrow{k_4} S_0 + M \qquad \text{phosphorescence quenching} \quad (4.17)$$

$$T_1 \xrightarrow{A''} S_0 + h\nu'' \qquad \text{phosphorescence} \qquad (4.18)$$

$$T_1 \xrightarrow{k_5} S_0 \qquad \text{ISC} \qquad (4.19)$$

Solution of the stationary state equations for $[S_1]$ and $[T_1]$ leads to the results

$$\phi_f = \frac{A'}{A' + k_2 + k_3 + k_1[M]} \qquad (4.20)$$

and

$$\phi_p = \phi_f \frac{A''}{A'} \cdot \left(\frac{k_2}{A'' + k_5 + k_4[M]} \right) \qquad (4.21)$$

Measurement of both ϕ_f and ϕ_p under conditions where bimolecular quenching is negligible (or, alternatively, extrapolation to $[M] = 0$) can thus lead to values for $(k_2 + k_3)$ and k_5/k_2 if A' and A'' are known. Direct measurement of the rates of process (4.15) have been made by chemical means (see the end of Section

5.7) in some cases, and the values of k_2 obtained suggest that $k_3 \ll k_2$; with k_2 known, k_5 may be estimated.

The discussion of the preceding paragraph indicates how kinetic data and measurements of quantum yields might ideally be used to assess the importance of the several photochemical processes (4.12–4.19). As we mentioned earlier in this chapter, quantum yields, especially for phosphorescence, are not known with precision in many instances and doubt still exists about the relative rates of, say, internal conversion and intersystem crossing. However, considerable progress in the accurate determination of quantum yields is being made at present, and the evaluation of individual rate constants by the methods outlined is the subject of much research.

It is in the nature of steady state kinetic calculations that *ratios* of rate constants are obtained: for example, the expressions for the intensities or quantum yields in Eqs. (4.6), (4.10), (4.11), (4.20) and (4.21) all involve ratios of rate constants to the Einstein A factor for emission. Individual rate constants can often be determined from a comparison of kinetic data obtained under stationary conditions with those obtained under non-stationary conditions. For the present purposes, the non-stationary experiment often involves determination of fluorescence or phosphorescence life-times (τ_f, τ_p). If a process follows first-order kinetics described by a rate constant k, the mean lifetime, τ (the time taken for the reactant concentration to fall to $1/e$ of its initial value), is given by

$$\tau = 1/k \tag{4.22}$$

If the loss of luminescent species after the exciting radiation is shut off is unimolecular or pseudo-first-order, then we may define mean lifetimes for the decay of emission as the inverse of the sum of all the effective first-order rate constants. Thus, for the very simple reaction scheme (4.2)–(4.4)

$$\tau = (A + k_q[M])^{-1} \tag{4.23}$$

Measurement of τ as a function of $[M]$ and extrapolation to $[M] = 0$ then yields a value for A; since A can also be calculated, via B, from extinction coefficients in absorption, the measured value may afford a check on the calculated one (cf. Section 4.6). In general, when $[M] \neq 0$, the observed lifetime is shorter than the 'natural' radiative lifetime ($= 1/A$). Where intramolecular loss processes occur, and the stationary-state kinetics are described by Eq. (4.11), then

$$\tau = (A + k_1 + k_q[\text{M}])^{-1} \qquad (4.24)$$

If a reliable value of A may be calculated from B, then k_1 may be determined explicitly from τ without the need for absolute quantum yield determinations. Many quoted rate constants for IC or ISC processes derive, in fact, from lifetime measurements.

4.3   FLUORESCENCE

An electronically excited atom must lose its energy either by emission of radiation or by collisional deactivation: chemical decomposition is not possible, and radiationless degradation (involving an increase in translational energy) is extremely improbable. At low enough pressures, therefore, fluorescent emission is expected from all atoms. Many molecular species, however, either do not exhibit fluorescence or fluoresce weakly even when bimolecular reaction or physical deactivation does not occur. Some general principles can suggest whether a polyatomic organic molecule is likely to be strongly fluorescent. First, absorption must occur at a wavelength long enough to ensure that chemical dissociation does not take place. Absorption to an unstable state is clearly very unlikely to result in fluorescence. Further, in many molecules in which the absorption maximum corresponds to an energy greater than the cleavage energy of the least stable bond no fluorescence is observed. Secondly, intramolecular energy transfer must be relatively slow compared to the rate of radiation. This appears to mean that ISC from $S_1 \rightsquigarrow T_1$ must be slow (we have mentioned in Section 4.2 and shall discuss later in this section the inefficiency of the IC $S_1 \rightsquigarrow S_0$ process); in Section 4.5 we shall see that ISC is normally slower for $^1(\pi, \pi^*) \rightsquigarrow {}^3(\pi, \pi^*)$ states than for ISC involving n, π^* states, and that the efficiency of the process increases as the energy separation of S_1 and T_1 decreases. Experimental observations of fluorescence are in accord with these ideas: the simpler carbonyl compounds, in which the longest absorption corresponds to an $n \rightarrow \pi^*$ transition, are rarely fluorescent (but often phosphorescent) while aromatic hydrocarbons ($\pi \rightarrow \pi^*$ absorption) are frequently fluorescent. Increasing conjugation in hydrocarbons shifts the first ($\pi \rightarrow \pi^*$) absorption maximum towards longer wavelengths, and thus increases the probability of fluorescence rather than decomposition. High ring density of the π electrons also seems important for high fluorescence yields. Geometrical factors such as rigidity and planarity also affect the efficiency of fluorescence. For example, ϕ_f for fluorene,

in hexane is about 0·54 while for biphenyl,

$\phi_f = 0.23$.

The simplest type of fluorescence is *resonance fluorescence*, in which the radiation emitted is of the same wavelength as the exciting radiation. Resonance fluorescence is observed only in the gas phase at low pressures, and only with atoms or simple molecules. For example, in I_2 vapour, at a pressure around 10^{-2} torr, resonance fluorescence is observed on excitation by light of suitable wavelength. Absorption of monochromatic light, of wavelength corresponding to a specific vibrational transition $(v', 0)$ in the $^3\Pi \leftarrow {}^1\Sigma_g^+$ system of I_2, populates exclusively the v' level of the upper state and radiation from that state gives rise to the resonance fluorescence (see Fig. 3.1). Transitions also occur from v' to v'' levels higher than zero, so that a progression of bands is observed at wavelengths longer than the exciting wavelength,† in accordance with an empirical observation of Stokes: the lines are called *Stokes lines*. (Much weaker *anti-Stokes lines* are observed at shorter wavelengths than the exciting wavelength, and result from absorption from $v'' > 0$ and fluorescence to lower v'' levels: thermal populations of $v'' > 0$ are small and the anti-Stokes lines are correspondingly weak.) Irradiation by polychromatic light can obviously excite many v' levels, and fluorescent emission can then be observed from all these levels, up to the dissociation limit (in I_2, up to $\lambda = 499$ nm).

Radiationless transitions can lead to a rapid depopulation of v' levels near the crossing point of potential energy curves: this depopulation is one of the reasons why resonance fluorescence of complex molecules is rare, even at low pressures. As was pointed

---

†The bands are often referred to as part of the resonance emission spectrum since they derive from the level of v' initially populated. Properly speaking, however, the resonance line is the same wavelength as the exciting line.

out in Section 3.2, if the crossing process occurs only ten times more rapidly than radiation (e.g. if, for an ordinary 'allowed' fluorescence, the rate constant for crossing is about $10^9 \, \text{s}^{-1}$), then the emitted intensity will be reduced by a factor of about ten, and 'breaking-off' in emission is a sensitive test of predissociation. Nitrogen dioxide fluorescence illustrates this phenomenon nicely. We have noted in Section 3.3 (cf. Fig. 3.6) that the primary quantum yield for decomposition of NO_2 increases sharply at wavelengths shorter than those at which the absorption spectrum becomes diffuse ($\lambda < 400$ nm approximately). The quantum yield for fluorescence in NO_2 is intense for excitation at $\lambda = 435 \cdot 8$ nm, weak at $\lambda = 404 \cdot 7$ nm and not observed at $\lambda = 365 \cdot 0$ nm. Electronically excited NO_2 can also be formed chemically by the reaction

$$O + NO + M \longrightarrow NO_2^* + M \qquad (4.25)$$

(see Section 4.7), and the short wavelength limit of chemiluminescent emission from reaction (4.25) corresponds exactly to the wavelength of onset of diffuseness in absorption.

Stepwise collisional relaxation of vibrational excitation is a relatively efficient process, cross-sections for single-quantum deactivation being between 1 % and 100 % of the gas-kinetic cross-section for many quenching gases. Resonance fluorescence is not expected, therefore, at pressures at which the kinetic collision frequency greatly exceeds the spontaneous emission rate, and, for $A \sim 10^8 \, \text{s}^{-1}$, this confines observation of resonance emission to pressures at least below about 1 mm Hg (and less, if A is smaller than $10^8 \, \text{s}^{-1}$). Lower vibrational levels of the upper electronic state are populated from the v' level produced on absorption, and at moderate pressures, at which emission and vibrational quenching still compete, emission may be observed from all vibrational levels in the upper state up to v'. For example, although individual bands are not seen, the fluorescence of NO_2 shifts to progressively longer wavelengths as the total pressure in the system is increased. The quenching data for NO_2 fluorescence given in Table 4.1 refer to relatively narrow-band emission selected from the total fluorescence spectrum, and at least part of the quenching is of vibrational rather than electronic emission. It must be emphasised that this is an unusual situation, and the example is chosen to illustrate certain features. Vibrational quenching is normally considerably more efficient than electronic quenching, and this is reflected in the relatively large ratio $\pi\sigma_q^2/\pi\sigma_{coll}^2$ (Table 4.1). Further, the value of k_q might be expected to depend on the v' level from which emission is observed, and to be smaller for longer-wavelength emission, since low-lying vibrational levels may be populated from higher levels as well as depopulated in the quenching processes. Table 4.2 shows values of k_q with $M = NO_2$ and $M = Ar$ for fluorescence at several wavelengths (calculated on the basis of

Table 4.2. Rate constants for quenching of NO_2 fluorescence as a function of wavelength of emission. Data of G. H. Myers, D. M. Silver and F. Kaufman, *J. chem. Phys.* **44,** 718 (1966)

Wavelength of fluorescence (nm)	$10^{-10} k_q(\text{l mol}^{-1}\text{s}^{-1})$		$\dfrac{k_q(M = Ar)}{k_q(M = NO_2)}$
	$M = NO_2$	$M = Ar$	
458·0	6·53	—	—
468·0	6·67	2·59	0·39
489·0	5·30	1·93	0·36
548·0	3·38	1·21	0·36
580·0	2·50	—	—
592·0	—	0·92	—
633·0	1·96	0·72	0·37

Excitation wavelength = 435·8 nm; A assumed to be $2.2 \times 10^4 \text{s}^{-1}$

$A = 2.2 \times 10^4 \text{ s}^{-1}$). Not only does k_q decrease with increasing wavelength of emission for both gases, but the ratio of quenching efficiencies of the two gases also remains virtually constant over the range, thus implying a simple mechanism in which there is a single radiative lifetime and in which k_q is independent of v'.

At higher gas pressures, at which the collision rate greatly exceeds the rate of emission, vibrational relaxation is essentially complete, and no fluorescence is observed from $v' > 0$. Vibrational relaxation is extremely probable in solution, and fluorescence from vibrationally excited levels is never observed in the liquid phase. Furthermore, neither the fluorescence spectrum nor deactivation rates are affected by changes in exciting wavelength so long as it lies within the absorption bands. $S_0 \rightarrow S_1$ transitions in organic compounds are often partially forbidden; to obtain sufficient light absorption to render gas-phase fluorescence detectable frequently requires large pressures, which result in vibrational relaxation to $v' = 0$. This relaxation, together with the probability of radiationless loss in complex species, accounts for the rarity of resonance or vibrationally hot emission phenomena in organic molecules.

The intensity of each vibrational emission band depends, in the same way as absorption intensities, on the operation of the Franck–Condon principle. Simple diatomic species frequently have greatly different internuclear separations in ground and excited state: Fig. 2.4 (b) is representative of the type of situation for O_2, where the upper state is larger than the lower state. Absorption in O_2 originates almost entirely from $v'' = 0$, and is most probable to v' levels from 7 to 11, and $\lambda_{max} \sim 185$ nm, while fluorescence, at pressures at which vibrational relaxation is complete, is strongest around the 0,14) band at $\lambda \sim 340$ nm. In contrast, the (0,0) band is the most intense for many organic molecules, and the maxima of intensity both in absorption and in emission therefore correspond to the same transition. This suggests that upper and lower electronic states of such molecules must be of similar size and shape, and it is likely that the vibrational *spacings* will be the same in both states. Figure 4.3 shows absorption and emission spectra of a

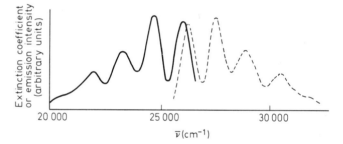

Figure 4.3. Absorption and emission spectra of a solution of anthracene in benzene, plotted on a frequency scale to reveal the energy spacing of vibrational bands. The absorption spectrum is shown as a solid line, and the emission spectrum as a broken line. (Data of E. J. Bowen (Ed.), Luminescence in chemistry, *p. 10, Van Nostrand, Princeton, N.J. (1968))*

solution of anthracene in benzene. The two spectra are almost mirror images of each other on the frequency (i.e. energy), rather than wavelength, scale employed. Figure 4.4 illustrates the energy levels in upper and lower states: because spacings are similar, the (0, 1) emission band will lie at the same energy below the (0, 0) band as the (1, 0) absorption band lies above it, and so on. This 'mirror-image' relationship is of frequent occurrence in the fluorescence of organic substances; assumption of its existence is sometimes useful in sorting out overlapping emission spectra. Whether or not there is a mirror image, the spacing of emission bands indicates the vibrational levels in the ground electronic state, while the spacing of absorption bands depends on vibrational spacing in the upper state.

The (0, 0) bands in Fig. 4.3 are seen to lie at slightly different wavelengths in absorption and in emission; Fig. 4.5 shows a more pronounced separation of (0, 0) bands for dimethylnaphtheurhodine

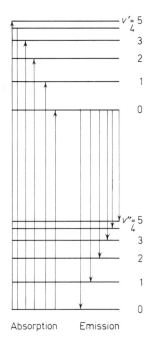

Absorption   Emission

Figure 4.4. Two electronic states with similar vibrational spacings; the absorption and emission bands bear a mirror-image relationship to each other

in two solvents. The separations are caused by energy loss to the solvent environment. The equilibrium interactions with the solvent may be different for ground and excited states of the solute (these are mainly electrical interactions, via the dipole moment of the solute if sizes are similar in both states). Although the species cannot relax to the equilibrium interaction energy during the absorption process (i.e. in about 10^{-15} s), it can do so before fluorescent emission occurs (i.e. in about 10^{-8} s). Thus the energy of 'equilibrium $v' = 0$' is rather lower than that of $v' = 0$ populated by absorption, and there is separation of (0, 0) bands. The magnitude of the separation depends on the dipole moment of the excited state of the emitting species, and also on the polarity of the solvent (cf. Fig. 4.5 for hexane and ethanol as solvents). Measurement of (0, 0) band separations is, in fact, used to estimate dipole moments of excited species. At very low temperatures the separations become small because molecular movements are frozen, and the separations may also be small at very high temperatures because of violent motions.

Organic fluorescence almost always originates from the *lowest excited singlet* level, S_1, even though absorption may initially

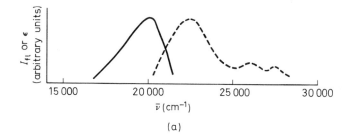

(a)

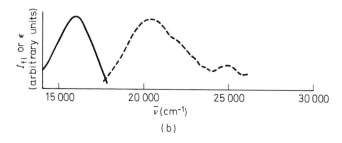

(b)

Figure 4.5. Absorption (solid line) and fluorescence (broken line) bands for solutions of dimethylnaphtheurhodine in (a) hexane, (b) ethanol. There is a pronounced separation of the (0,0) bands (intensity maxima). (Data of E. J. Bowen (Ed.), Luminescence in chemistry, *p. 10, Van Nostrand, Princeton, N.J. (1968))*

populate a higher singlet (e.g. S_2, S_3, S_x, etc). Apparently there is rapid internal conversion from S_x to S_1^v followed by vibrational degradation (possibly the process involves intersystem crossings via intermediate triplet states). Internal conversion $S_1 \rightsquigarrow S_0$ must be much slower than $S_x \rightsquigarrow S_1$, since emission from S_1 can compete with radiationless loss and, perhaps more tellingly, since forbidden intersystem crossing $S_1 \rightsquigarrow T_1$ can compete and lead to phosphorescent emission. Some recent studies of Ermolaev have suggested that in rigid glasses at low temperatures the $S_1 \rightarrow S_0$ process occurs only via triplet states:

$$S_1 \overset{\text{ISC}}{\rightsquigarrow} T_1 \overset{\text{ISC}}{\rightsquigarrow} S_0 \qquad (4.26)$$

However, measurements of $\phi_{\text{ISC}(S_1 \rightsquigarrow T_1)}$ and of ϕ_f for many aromatic molecules in hydrocarbon solution suggest that direct internal conversion $S_1 \rightsquigarrow S_0$ must occur at room temperatures in fluid solution. Table 4.3 shows values of ϕ_{ISC} and ϕ_f for benzene,

Table 4.3. Quantum yields for intersystem crossing $(S_1 \rightsquigarrow T_1)$ and fluorescence $(S_1 \rightarrow S_0 + h\nu)$. Intersystem crossing data from J. G. Calvert and J. N. Pitts, Jr., *Photochemistry*, Table 4–17, Wiley, New York (1966); fluorescence data from E. J. Bowen, *Adv. Photochem.* **1**, 23, Table III (1963)

Compound	ϕ_{ISC} (in benzene)	ϕ_f (in hexane)	$1 - \phi_{\text{ISC}} - \phi_f$
Benzene	0·24	0·04	0·72
Naphthalene	0·39	0·10	0·51
Fluorene	0·31	0·54	0·15
Acenaphthalene	0·47	0·31	0·22

naphthalene, fluorene and acenaphthalene in hydrocarbon solvents at room temperatures. If no *collisional* deactivation of S_1 occurs, and if all radiationless loss of S_1 occurs via T_1, then $\phi_{\text{ISC}} + \phi_f$ should be unity. Table 4.3 shows that $\phi_{\text{ISC}} + \phi_f \neq 1$, and the discrepancy is considerable for the simpler hydrocarbons. Thus, even though some bimolecular quenching of S_1 probably does take place, it seems that internal conversion is significant in the liquid phase. However, we would reiterate the statement that, since the IC process is *competitive* with radiation or ISC, it is *relatively* slow. We shall see, in Section 4.5, that the efficiency of intramolecular energy transfer usually decreases as the energy difference between two levels increases, and it is possible that the inefficiency of the $S_1 \rightsquigarrow S_0$ process, especially compared with that of $S_x \rightsquigarrow S_1$, reflects the relatively large energy separation between S_1 and S_0.

The observation that fluorescent emission occurs only from S_1, and not from higher singlet states, is usual in organic photochemistry, and Kasha has enunciated a rule that *the emitting electronic level of a given multiplicity is the lowest excited level of that multiplicity.* Indeed, the only organic compound whose fluorescence appears to be an exception to this rule is azulene. However, the rule certainly does not hold strictly for atoms, with which radiationless energy loss can occur only in collisions, or

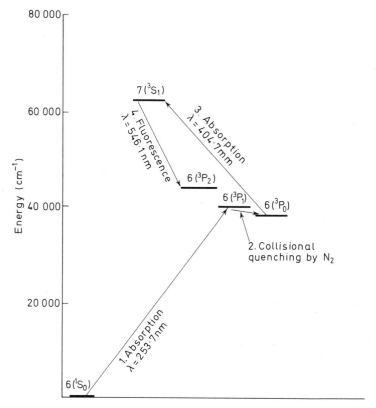

Figure 4.6. Energy levels for some atomic states of mercury. The stages involved in the excitation of green fluorescence ($\lambda = 546 \cdot 1$ nm) are shown

simple inorganic molecules, with which non-radiative intramolecular energy transfer is relatively infrequent. Perhaps the simplest example is the resonance 'fluorescence' of mercury atoms excited by the $\lambda = 253 \cdot 7$ nm line.† Emission (and absorption) are of the

---

†Fluorescence is probably the best description of the emission process even though it involves a change of multiplicity: the radiative lifetime is 'short'.

Hg $6(^3P_1) \rightarrow 6(^1S_0)$ line, and the emitting $^3P_1$ state lies *above* the $^3P_0$ state. Admittedly, in this example the states differ only in respect of the spin–orbit coupling, and the $^3P_0$ state cannot radiate to $^1S_0$ rapidly because of the $J = 0 \nrightarrow J = 0$ selection rule. Another elegant experiment involving mercury shows fluorescent radiation from an upper triplet state in which the principal quantum

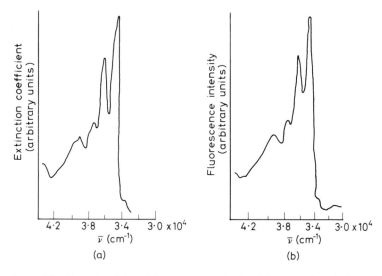

*Figure 4.7. Absorption (a) and fluorescence excitation (b) spectra for solutions of 1,2-benzanthracene in ethanol. (Data of C. A. Parker, Nature **182**, 1002 (1958))*

number is higher. Small amounts of N_2 can quench the $6(^3P_1)$ state to $6(^3P_0)$; green fluorescence at $\lambda = 546.1$ nm can then be excited by the $\lambda = 404.7$ nm line. Figure 4.6 shows the levels of Hg which are involved, and indicates the sequence of events leading to emission of the green line; emission clearly *does not* come from the lowest excited level of triplet multiplicity. If radiationless loss of energy cannot occur, then molecules may also emit from higher levels. Thus resonance phosphorescence from the $^1\Sigma_g^+$ state of O_2 is observed in the atmosphere, even though the $^1\Delta_g$ state (i.e. same multiplicity) lies below it. The potential energy curves for the two states do not cross, and intramolecular energy transfer cannot occur efficiently; further, radiationless transition is also forbidden by the angular momentum rule, since $\Delta\Lambda$ would be two.

Because fluorescent emission in organic compounds comes predominantly from the lowest vibrational level of the lowest excited singlet state of the molecule, it is often found that the fluorescence quantum yield is independent of the wavelength of exciting radiation. Since

$$I_f = I_{abs}\phi_f = I_0\phi_f(1 - e^{-\alpha cd}) \qquad (4.27)$$

at very low concentrations, where $\alpha cd \ll 1$,

$$I_f = I_0\phi_f\alpha cd \qquad (4.28)$$

It follows from the constancy of ϕ_f with exciting wavelength that I_f is proportional to α at any wavelength if the incident intensity is the same at all wavelengths. That is, the *fluorescence excitation spectrum* is the same as the absorption spectrum in sufficiently dilute solutions. This result forms the basis of *spectrofluorimetry*. Figure 4.7 shows part of the absorption spectrum (a) of 1,2-benzanthracene in ethanol and the fluorescence excitation spectrum (b) of a much more dilute solution. The technique clearly makes it possible to obtain 'absorption' spectra at very low solute concentrations (typically, at concentrations as low as 10^{-9} mol l^{-1}), and the great sensitivity of spectrofluorimetry is making it a useful analytical tool. It should be noted that the structure of a fluorescence *emission* spectrum, as distinct from the *excitation* spectrum, is often not sufficiently specific in condensed phases for molecular identification to be possible.

4.4 PHOSPHORESCENCE

Organic molecules trapped in rigid glassy media have long been known often to show a long-lived phosphorescent afterglow following irradiation by light. It is now understood that phosphorescence in organic molecules is emission of a 'forbidden' band, usually from a triplet level. Because of the long radiative lifetime of such transitions, collisional deactivation of the triplet competes effectively with radiation, and visible phosphorescence is not normally observed unless the collisional deactivation rate is sufficiently reduced. In rigid media species are unable to diffuse towards each other, and bimolecular deactivation is slow. The earliest investigations of phosphorescence employed solutions of dyes in gelatin, and subsequently in boric acid glass at room temperature. More satisfactory rigid media are now used: mixtures of ether, isopentane and ethanol (EPA) frozen at liquid nitrogen temperature (77K) are frequently employed, and thin films of various plastics are becoming popular as rigid matrices. The highest purity of the solvents is necessary to avoid swamping the phosphorescence of the solute by luminescence of the impurities.

Although the first observations of phosphorescence were confined to rigid glasses, it was soon appreciated that phosphorescence could appear in other phases. Emission from biacetyl vapour is one of the best-known examples of gas-phase phosphorescence. The development of high-sensitivity radiation detectors such as photomultipliers has allowed observation of very weak emission even in fluid solutions of phosphorescent substances. The luminescence is still of the phosphorescence system, although it is not long-lived, since the decay rate is determined by the rate of physical quenching.

Confirmation that the emitting species in phosphorescent organic molecules is a triplet has come from several sources. In the 1940s it was discovered that a solution of fluorescein in boric acid glass became paramagnetic under intense irradiation; more recently it has been shown that the paramagnetism and the phosphorescence decay at identical rates when irradiation ceases. The electron paramagnetic resonance (EPR) technique should be capable of detecting triplet species, and, although preliminary investigations were unsuccessful, triplet states of several molecules have been observed by EPR. The first unambiguous detection of a triplet by EPR was of the $\Delta M = \pm 1$ transition in an irradiated single crystal of naphthalene in durene; $\Delta M = \pm 2$ transitions have also been observed in irradiated naphthalene. Triplet concentrations, measured by EPR, in solid solutions of certain phosphorescent aromatic ketones decay, after irradiation, at the same rate as the phosphorescence.

Optical absorption to a higher triplet has afforded further evidence that the emitting state in phosphorescence is a triplet. Intense irradiation of a boric acid glass containing fluorescein leads to the appearance of a new absorption band due to triplet–triplet absorption. Flash photolysis (see Chapter 7), in which a sample is exposed to a brief, intense flash of light, has been used to produce high transient concentrations of triplet species: kinetic absorption spectroscopy of the system enables the build-up and decay of several singlet and triplet levels to be followed as a function of time.

We must now give some consideration to the nature of the

forbidden triplet–singlet emission. In Section 2.7 we suggested that electric dipole transitions could occur with $\Delta S \neq 0$ if S did not offer a good description of the system. In effect, this means that optical transitions between triplet and singlet states can take place if the triplet has some singlet character, or vice versa. In the organic molecules some 'mixing' of singlet and triplet states takes place as a result of a small amount of spin–orbit interaction. It turns out that the spin–orbit perturbation is forbidden between states of the same configuration, so that, for example, a $^3(\pi,\pi^*)$ state must 'borrow' its singlet character from $^1(n,\pi^*)$ and $^1(\sigma,\pi^*)$ states rather than from $^1(\pi,\pi^*)$. Similarly, a $^3(n,\pi^*)$ state mixes with a $^1(\pi,\pi^*)$ state. Since radiative transitions from $^1(\pi,\pi^*)$ states to the ground state are fully allowed, while transitions from $^1(n,\pi^*)$ are, in general, somewhat forbidden, it follows that $T(n,\pi^*) \rightarrow S_0$ transitions are more allowed than $T(\pi,\pi^*) \rightarrow S_0$. Thus the relative probability of triplet–singlet transitions from (n,π^*) and (π,π^*) states is opposite to that observed for singlet–singlet transitions. Experimental determinations of natural phosphorescence lifetimes agree with the predictions: in aromatic hydrocarbons having a $^3(\pi,\pi^*)$ state for T_1 the radiative lifetime is roughly 10–100 s, while for carbonyl compounds possessing a lowest triplet state of $^3(n,\pi^*)$ character the lifetime is usually 10^{-3}–10^{-2} s.

Absorption leading to direct population of an excited triplet state from the singlet ground state is weak because the transition is forbidden: decadic molar extinction coefficients may be as low as 10^{-5}. However, in some cases it has proved possible to excite phosphorescence by irradiation with light absorbed in the $T_1 \leftarrow S_0$ system. Just as with fluorescence, there is often a mirror-image relationship between absorption and phosphorescence spectra. It would appear, therefore, that in the relatively large organic molecules the vibrational spacings are nearly identical in all three lowest states (S_0, T_1 and S_1). The (0,0) separations in the $T_1 \leftarrow S_0$ absorption and emission spectra are, however, relatively large ($\sim 500\, cm^{-1}$) as a result of slight conformational differences between ground and excited states. Hence, triplet energies based solely on presumed (0,0) band maxima in emission only partly represent the energetics of an absorbing system. Good absorption spectra of $T_1 \leftarrow S_0$ transitions are difficult to obtain by ordinary techniques, but the weakness of the absorption makes it possible to use the *phosphorescence excitation spectrum* to determine the absorption spectrum (this is spectrophosphorimetry – cf. Section 4.3 for spectrofluorimetry).

Phosphorescence most commonly follows population of T_1 via intersystem crossing from S_1, itself excited by absorption of light. The T_1 state is usually of lower energy than S_1, and the long-lived (phosphorescent) emission is almost always of longer wavelength than the short-lived (fluorescent) emission. In a very few cases it is claimed that strong coupling between excited states leads to S_1 lying *below* T_1, and normal phosphorescence will not be observed in molecules displaying this behaviour. For example, species possessing an (n,π^*) state for S_1 frequently show high phosphorescence efficiencies, but 9,10-diazophenanthrene fluoresces, since $^1(n,\pi^*)$ is probably below $^3(\pi,\pi^*)$. The relative importance of fluorescence and phosphorescence depends on the rates of radiation and intersystem crossing from S_1; the absolute quantum yields depend also on intermolecular and intramolecular energy loss processes, and phosphorescent emission competes not only with collisional quenching of T_1, but also with intersystem crossing to S_0. The difference between the overall rate of triplet production from S_1 and the rate of phosphorescent emission can be used to calculate the efficiency of the $T_1 \rightsquigarrow S_0$ process under conditions in which bimolecular quenching is negligible.

4.5 INTRAMOLECULAR ENERGY TRANSFER IN COMPLEX MOLECULES (2)

Now that the fundamental principles of fluorescent and phosphorescent emission have been given, the general topic of intramolecular energy transfer can be treated in greater detail.

Selection rules for radiationless transitions in simple molecules have been given in Section 3.3 in the discussion of predissociation. It is more difficult to develop similar rules for complex species, but at least the spin rule, $\Delta S = 0$, is still of importance. As for emission processes, the occurrence of transitions with $\Delta S \neq 0$ is a result of spin–orbit coupling in the molecule, and the transition probabilities for intersystem crossing follow virtually the same pattern as those discussed in the last section for radiative processes. On the basis of these ideas, El-Sayed has suggested the following 'rules' for spin-forbidden intramolecular energy transfer:

$$^{1 \text{ or } 3}(n, \pi^*) \leftrightarrow {}^{3 \text{ or } 1}(\pi, \pi^*); \quad {}^3(n, \pi^*) \nleftrightarrow {}^1(n, \pi^*);$$
$$^3(\pi, \pi^*) \nleftrightarrow {}^1(\pi, \pi^*) \tag{4.29}$$

Experimental evidence, to be given later, shows that the transition probability for energy transfer is some inverse function of the energy gap between the two states. The total electronic + vibrational energy must clearly be the same in both states, as otherwise excess energy would have to be converted to translation. Vibrationally excited levels of lower electronic states are therefore populated, and vibrational relaxation can subsequently occur. The operation of the Franck–Condon principle in radiationless transitions can account for the increase in crossing rate with decrease in ΔE. If the geometries of two states X and Y are similar, the overlap of vibrational wave functions in the upper state X (with $v' = 0$) and the lower state Y (in vibrational level v'') will be greater for smaller v'' because of the oscillatory nature of the higher vibrational wave functions (see Section 2.8, and Figs. 2.3 and 2.4a). Thus the efficiency of crossing will increase as $v'' \to 0$; that is, crossing to a state will be favoured if the state can be populated near $v'' = 0$, which means that the *electronic* energy gap must itself be small. This point is nicely illustrated by the effect of deuteration of some aromatic hydrocarbons on the rate of ISC($T_1 \rightsquigarrow S_0$). Deuteration lowers C—H stretching frequencies, so that for a given electronic energy gap $\Delta E(T_1 - S_0)$, v'' in S_0 will be higher. Thus, in biphenyl the rate constant, k_t, for the ISC process is $2.9 \times 10^{-1}\,\mathrm{s}^{-1}$, but in decadeuterobiphenyl $k_t = 5.1 \times 10^{-2}\,\mathrm{s}^{-1}$. Again, $k_t = 4.2 \times 10^{-1}\,\mathrm{s}^{-1}$ in naphthalene, but only $9.0 \times 10^{-2}\,\mathrm{s}^{-1}$ in octadeuteronaphthalene. In neither case is the rate of radiative transition from T_1 to S_0 affected.

Cascade degradation through the vibrational levels of several intermediate electronic states possibly accounts for the high efficiency of some energy exchanges in which the energy gap is relatively large. Thus it is suggested that the intersystem crossing from $S_1 \rightsquigarrow T_1$, which leads to phosphorescence, is fast because it proceeds via a series of triplets lying above T_1 but below S_1. Until recently, there was no other evidence for the existence of these triplets. However, the second triplets of benzene and anthracene and of phenyl ketones have now been found to lie just below the first excited singlets.

The effect that the energy gap has on the rate of radiationless transition is illustrated in Fig. 4.8 for the $T_1 \rightsquigarrow S_0$ ISC in several aromatic hydrocarbons (in rigid solvents at 77 K). The electronic states involved are similar for each hydrocarbon, and the increase in rate of radiationless transition in going from benzene to anthracene follows the decrease in $\Delta E(T_1 - S_0)$. As a general rule, the larger the molecule, the smaller $\Delta E(T_1-S_0)$, and ϕ_p is correspondingly low in large molecules. The differences in the photochemistry of species whose absorption system is predominantly $\pi-\pi^*$ or $n-\pi^*$ are at least in part a result of the increased likelihood of $S_1 \rightsquigarrow T_1$ ISC for $S_1 = {}^1(n, \pi^*)$.† Three factors work towards this

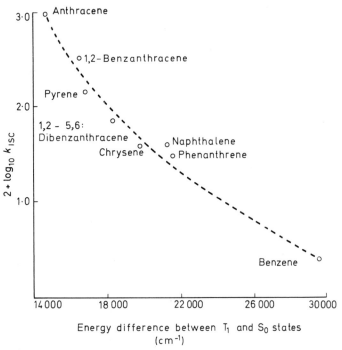

Figure 4.8. Dependence of rate constant for intersystem crossing (k_{ISC}) on energy gap between triplet (T_1) and ground states (S_0) for several species. The units for the rate constants are s^{-1}. (Data derived from J. G. Calvert and J. N. Pitts, Jr., Photochemistry, *Table 4–14, p. 301, Wiley, New York (1966))*

increased likelihood of radiationless transition for ${}^1(n, \pi^*)$ states. First, both radiative and radiationless transitions to the S_0 ground state are partially forbidden, so that the ${}^1(n, \pi^*)$ state survives

†$\phi_{ISC} \sim 1$ at room temperature for $S_1 \rightsquigarrow T_1$ in several aromatic ketones (benzene solution); for hydrocarbons the efficiency is much less (e.g. $\phi_{ISC} = 0.24$ for benzene).

longer than a $^1(\pi, \pi*)$ state and has more chance of undergoing ISC to T_1. Secondly, T_1 *may* be $^3(\pi, \pi*)$, to which ISC from $^1(n, \pi*)$ is favoured. Thirdly, the energy gap, $\Delta E(S_1 \to T_1)$ is often small compared to the gap from $^1(\pi, \pi*)$ states. The splitting for singlet to triplet $(\pi, \pi*)$ is often from about $10\,000\,\text{cm}^{-1}$ to $20\,000\,\text{cm}^{-1}$, while that for $(n, \pi*)$ states is usually only $1500\,\text{cm}^{-1}$– $5000\,\text{cm}^{-1}$. There are six possible arrangements of the energy levels where both $(\pi, \pi*)$ and $(n, \pi*)$ states exist, and Fig. 4.9 shows the types of species to which these arrangements apply. Note that in cases such as (a), where S_1 and T_1 are both $(n, \pi*)$ states, T_1 may be populated via the favourable ISC transition to $^3(\pi, \pi*)$. In many carbonyl compounds phosphorescence is strong, and fluorescence weak or non-existent (typically $\phi_p/\phi_f > 1000$) because of this efficient crossing from S_1 to T_1; a fairly large fraction of the molecules reaching T_1 phosphoresce, the magnitude of the $T_1 \to S_0$ gap partly determining the rate of radiationless, $T_1 \leadsto S_0$ ISC, decay. The large $\Delta E(S_1 \to T_1)$ and the unavailability of $^1(n, \pi*) \to {}^3(\pi, \pi*)$ transitions in hydrocarbons make $S_1 \leadsto T_1$ ISC less probable, and fluorescence becomes important as well as phosphorescence (and decay) from T_1. In connection with the importance of $\Delta E(S_1 \to T_1)$, it is noteworthy that triphenylene, which has an abnormally low ΔE $(5000\,\text{cm}^{-1})$ for a hydrocarbon, has $\phi_p/\phi_f = 2\cdot8$ in a glass at 77 K, while ϕ_{ISC} in benzene solution at room temperature is $0\cdot95$. These values should be compared with those for the more typical hydrocarbon, naphthalene: $\Delta E = 10\,500\,\text{cm}^{-1}$, $\phi_p/\phi_f = 0\cdot09$ (77 K), $\phi_{ISC} = 0\cdot39$ (benzene solution, room temperature). In general, the separation $\Delta E(S_1 - T_1)$ decreases with increasing molecular size for $T_1 = {}^3(\pi, \pi*)$, and ISC from S_1 to T_1 becomes more important. This result does *not* apply to the series benzene, naphthalene, anthracene.

Emitting triplet states have normally been characterised as $(n, \pi*)$ or $(\pi, \pi*)$ on the basis either of specific resolvable structure, e.g. of C=O vibrations, or of phosphorescence lifetime measurements, the $^3(\pi, \pi*)$ lifetimes being much greater than those of the $^3(n, \pi*)$ state. The value of $\Delta E(S_1 \to T_1)$ offers some indication of the natures of S_1 and T_1, although this kind of criterion must obviously be used with care. More recently, the influence of heavy atom environments (see later) on $T_1 \leftarrow S_0$ absorption intensities, or $T_1 \to S_0$ phosphorescence lifetimes, has been used to distinguish $(n, \pi*)$ and $(\pi, \pi*)$ states. Probabilities for transitions to or from $^3(\pi, \pi*)$ states are increased by the heavy atom environment, and the technique has been used to show that, although T_1 for aceto-

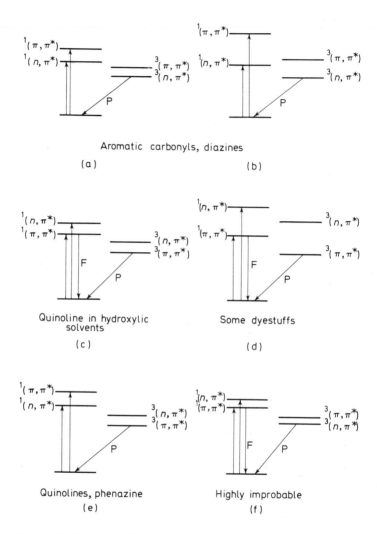

Figure 4.9. *The six possible arrangements of singlet and triplet (π,π^*) and (n,π^*) excited states; the diagram shows the types of molecule for which each arrangement is likely to apply.* (From F. Wilkinson and A. R. Horrocks in Luminescence in chemistry *(Ed. E. J. Bowen), p. 130, Van Nostrand, Princeton, N.J. (1966))*

phenone is $^3(n, \pi^*)$, for several substituted acetophenones, the state is $^3(\pi, \pi^*)$.

Table 4.4 summarises some data obtained by Ermolaev for emission in selected organic molecules, together with values for ϕ_{ISC}; all data, with the exception of those for ϕ_{ISC}, refer to experiments performed in an alcohol–ether glass at 77 K. As anticipated, ϕ_p/ϕ_f is large for molecules possessing small $\Delta E(S_1-T_1)$, especially if S_1 is an (n, π^*) state (e.g. in the carbonyl compounds). Both A_p and k_t, the rate constants for radiative and radiationless $T_1 \rightarrow S_0$ decay, respectively, are much greater for the favoured transitions with $T_1 = {}^3(n, \pi^*)$; although A_p varies over a range of 2×10^4, k_t/A_p shows a much smaller range. According to a simple scheme, in which T_1 is depopulated *only* by $T_1 \overset{ISC}{\rightsquigarrow} S_0$ and phosphorescence, the total quantum yield, equivalent to ϕ_{ISC}, should be equal to $\phi_p + (k_t/A_p)\phi_p$. The values of $\phi_p[1 + (k_t/A_p)]$ are given in Table 4.4, and compared with ϕ_{ISC} measured in solution. The calculated values of $\phi_{ISC(S_1 \rightsquigarrow T_1)}$ in the 77 K glass are generally higher than, or the same as, the measured ϕ_{ISC} for room temperature solutions; loss processes for $S_1 \rightarrow S_0$ (fluorescence, IC, and collisional deactivation) are presumably of greater importance in the liquid phase. The sum of fluorescence and calculated ISC $(S_1 \rightarrow T_1)$ yields, given as ϕ_t in Table 4.4, is near unity for all the species listed, suggesting that radiationless IC from, and collisional deactivation of, S_1 is negligible in the low-termperature glass.

Phase certainly appears to play an important part in determining the efficiency of the forbidden ISC processes. Phosphorescence yields are often small in solution, and ISC $(T_1 \rightsquigarrow S_0)$, together with collisional quenching, proceeds faster in the liquid phase.† The activation energy for radiationless decay is very low at low temperatures in viscous solutions (e.g. for ketones in a very viscous ester), but there is a sharp increase in activation energy above about 200 K when interactions with the solvent become significant. It is likely that the occurrence of molecular collisions perturbs the system sufficiently to enhance the probability of the spin-forbidden process.

---

†There has been some discussion about whether small ϕ_p in solution is really a result of quenching by traces of adventitious impurities (especially O_2) present in the solvent. This view has received support from the dramatic decrease in the published values for 'unimolecular' decay of the anthracene triplet as increasingly oxygen-free solvents have been employed. However, it has been shown that radiationless decay *does* still occur when the solvent is too viscous to allow enough diffusion for quenching to be important.

Table 4.4. Emission and radiationless transition data for some organic molecules (in alcohol–ether glass at 77K) arranged in ascending order of ϕ_p/ϕ_f. Except for ϕ_p/ϕ_f, the data are from V. L. Ermolaev, *Soviet Phys., Usp.* 333 (Nov.–Dec. 1963); ISC data are from J. G. Calvert and J. N. Pitts, Jr., *Photochemistry*, Table 4-17, Wiley, New York (1966)

Compound	ΔE (T_1-S_0) (cm^{-1})	ΔE (S_1-T_1) (cm^{-1})	Rate constant, k_r for $T_1 \overset{ISC}{\longrightarrow} S_0$ (s^{-1})	Transition probability, A_p for $T_1 \to S_0$ $+ h\nu$ (s^{-1})	k_r/A_p	ϕ_p	ϕ_f	ϕ_p/ϕ_f	$[1+ (k_r/A_p)]\phi_p$	ϕ_t^a	ϕ_{isc}^b
1-Methyl-naphthalene	21 000	10 450	4.5×10^{-1}	2.0×10^{-2}	22.5	0.02	0.43	0.05	0.47	1.00	0.48
Naphthalene	21 250	10 500	4.2×10^{-1}	1.6×10^{-2}	26.3	0.03	0.29	0.09	0.82	1.11	0.39
Carbazole	24 600	4 900	6.3×10^{-2}	6.9×10^{-2}	0.91	—	—	0.55	—	—	0.36
Phenanthrene	21 700	7 200	2.6×10^{-1}	4.6×10^{-2}	5.65	0.13	0.12	1.1	0.86	0.98	—
Quinoline	21 700	10 200	6.6×10^{-1}	7.7×10^{-2}	8.57	0.10	0.05	1.9	0.95	1.00	0.32
Triphenylamine	24 500	4 500	5.7×10^{-1}	8.6×10^{-1}	0.66	—	—	15	—	—	0.88
Acetophenone	25 750	1 750	1.7×10^2	2.8×10^2	0.61	0.62	0.00	>1000	1.00	1.00	0.99
Benzaldehyde	24 950	1 800	3.5×10^2	3.4×10^2	1.03	0.49	0.00	>1000	0.99	0.99	—

[a] $\phi_t = \phi_f + \phi_p[1+(k_r/A_p)]$.
[b] ϕ_{isc} results refer to benzene solutions at room temperature.

The perturbing influence of the environment on the rate of spin-forbidden transitions is seen in the effects of the addition of paramagnetic molecules to the solvent. Although O_2 and NO decrease phosphorescence yields because of their participation in efficient bimolecular quenching, they increase the rates both of optical transition and of ISC. Absorptions of the $T_1 \leftarrow S_0$ transition are also increased in intensity when the paramagnetic species is present, and, for example, the $T_1 \leftarrow S_0$ absorption in benzene ($\lambda \sim 310$–350 nm) practically disappears when the last traces of oxygen are removed. The most dramatic demonstration of the increase in $T \leftarrow S$ absorption is afforded by pyrene solutions, which are normally colourless but which turn deep red in the presence of high pressures of oxygen. Heavy atoms in an environment also increase the probability of $S \leftrightarrow T$ radiative and radiationless transitions by inducing appreciable spin–orbit coupling in the solute. Thus solutions of anthracene and some of its derivatives become less fluorescent on addition of bromobenzene, while the triplet–triplet absorption intensity increases as a result of enhanced $S_1 \rightsquigarrow T_1$ ISC. As we noted earlier, these effects are most significant for transitions involving (π,π^*) excited states. Spin–orbit coupling is almost negligible in symmetric aromatic compounds, and it is here that perturbation by the environment has relatively the largest effect; significant spin–orbit coupling already exists in (n,π^*) states, and the effect of external perturbation is less pronounced. The effects are noticed in both solid and fluid solutions; for example, the phosphorescence lifetime of benzene in glasses at 4·2 K decreases from 16 s in CH_4 or Ar to 1 s in Kr and to 0·07 s in Xe; at the same time, the ratio ϕ_p/ϕ_f increases, and all processes $S_1 \overset{\text{ISC}}{\rightsquigarrow} T_1$, $T_1 \rightarrow S_0 + h\nu$ and $T_1 \overset{\text{ISC}}{\rightsquigarrow} S_0$ may proceed more rapidly in the solvents of higher atomic weight.

*Intra*molecular perturbation of transition probabilities is also important. Table 4.5 shows A_p, k_t and ϕ_p/ϕ_f for naphthalene and some of its halo-derivatives. Substitution by one iodine atom increases the transition probability for optical emission (by a factor of nearly 10^4) and the rate of $T_1 \rightsquigarrow S_0$ ISC. Further, the increase in ϕ_p/ϕ_f results mainly from increased probability of $S_1 \rightsquigarrow T_1$ ISC in the substituted molecules. Similar effects are observed on substitution in many other species. Indeed, the major effect of substitution on the photochemistry of a species seems to lie not so much in changes in energy levels (the first triplet and excited singlet levels lie at 21 250 cm^{-1}, 31 750 cm^{-1}, respectively, in naphthalene, and shift only to 20 500 cm^{-1}, 31 000 cm^{-1} in 1-iodonaphthalene), as in changes in the relative probabilities of fluorescence, phos-phorescence and the IC and ISC processes.

Strong intramolecular perturbations may also arise when certain metal ions are chelated to an organic molecule. It is interesting that the natural porphyrins chlorophyll and haemin display markedly different photochemical behaviour: chlorophyll has diamagnetic

Table 4.5. Effect of halogen substitution in naphthalene on the rates of spin-for-bidden processes. Data from V. L. Ermolaev, *Soviet Phys., Usp.* 333 (Nov.–Dec. 1963)

Compound	$A_p(s^{-1})$ $T_1 \rightarrow S_0 + h\nu$	$k_t(s^{-1})$ ISC $T_1 \rightsquigarrow S_0$	$\phi_p\phi_f$
Naphthalene	0·016	0·42	0·09
1-Fluoronaphthalene	0·036	0·63	—
1-Chloronaphthalene	0·57	1·7	5·2
1-Bromonaphthalene	7·0	43	164
1-Iodonaphthalene	100	400	> 1000

magnesium as its central ion, while haemin has paramagnetic iron (see Fig. 8.7 for the structure of chlorophyll).

This section is concluded with a description of a rather different kind of intramolecular energy transfer process in which energy is transferred from one *part* of a molecule to another (rather than between different *states*). Irradiation of the compound

with light absorbed solely by the naphthalene group leads to emission of anthracene fluorescence, and singlet excitation is transferred even though the absorbing and emitting groups are separated by an 'insulating chain' of three CH_2 groups. Triplet–triplet spatial energy transfer is also known. The naphthylalkyl benzophenones absorb radiation at $\lambda \sim 366$ nm to excite exclusively the benzo-phenone like unit to its singlet state; ISC to the triplet state follows, and excitation is then transferred to the naphthalene nucleus, which emits its characteristic phosphorescence spectrum. The reaction is unlikely to proceed via singlet energy exchange followed by ISC in the naphthalene group, because S_1 for naphthalene is higher than that for benzophenone. Irradiation at $\lambda \sim 313$ nm ex-cites S_1 of the naphthalene moiety, and singlet excitation can be transferred to the benzophenone unit, whence it eventually returns to the naphthalene as triplet excitation. Energy transfer of this kind can also occur in some rare-earth chelates. For example, irradiation of a low-temperature solution of europium benzoyl-acetonate leads to emission of the 613 nm $Eu(^5D \rightarrow {}^7F)$ line. First the β-diketone ligand absorbs light to reach a singlet state, and ISC then populates the triplet, which passes on its energy to the euro-pium ion. Special interest in this process arises from the possibility of achieving population inversion and, hence, laser action (cf. Sections 2.3 and 5.6).

4.6 'DELAYED' FLUORESCENCE

In the earliest experiments on the phosphorescence of fluorescein in boric acid glass, it was recognised that at least two mechanisms were operating to give long-lived emission: the processes were called α- and β-phosphorescence. β-phosphorescence is the ordinary triplet–singlet emission described in previous sections, and its intensity is relatively insensitive to temperature. Several types of α-phosphorescence can be distinguished, and in this section we shall discuss that form which is sometimes known as E-type delayed fluorescence, after eosin, which displays the phenomenon (P-type delayed fluorescence, named after pyrene, will be mentioned in Section 5.5).

E-type delayed fluorescence spectra show features characteristic of the normal, short-lived, fluorescence. However, the emission decays at the same rate as phosphorescence; further, no emission is observed at low temperatures, and there is an activation energy for the process. It is now understood that this kind of delayed fluorescence arises from thermal activation of S_1 from $T_1(v = 0)$; the rate of activation is slow compared to the rate of loss of either T_1 or S_1, so that the decay of delayed fluorescence is determined by the decay of T_1. Figure 4.10 illustrates this excitation mechanism. The activation energy for the emission should be identical to $\Delta E(S_1-T_1)$, and Table 4.6 shows the remarkable agreement obtained for some molecules between spectroscopic energy differences

Table 4.6. Activation energies, E_a, for E-type delayed fluorescence and $\Delta E(S_1 - T_1)$ obtained spectroscopically[a]. Data from F. Wilkinson and A. R. Horrocks, in *Luminescence in chemistry*, Table 7.7, p. 148, Van Nostrand, London (1968).

System	$\Delta E(S_1 - T_1)$ (kcal/mol)	E_a (kcal/mol)
Eosin in ethanol	10·9	9·7
Eosin in glycerol	10·5	10·2
Fluorescein in boric acid	9·2	8±1
Proflavine in glycerol	8	8
Acriflavine in glucose	7·5±0·5	8·0±0·5
Acriflavine dimer in glucose	5·0±0·5	5·5±0·5
Acriflavine adsorbed on silica gel	5	8

[a]Original source gives energies in calorie units: 1 cal = 4·184 J.

and experimental activation energies (calculated from the variation of emission intensity with temperature).

In solution rapid vibrational degradation after ISC from S_1^0 to T_1^v means that re-excitation must be from the $v = 0$ level of T_1. At low pressures in the gas phase, this need not necessarily be the case, and reversible intramolecular energy exchange can occur in which thermal activation energy is not necessary. It is suggested that the ground and first excited electronic states of NO_2 are in rapid equilibrium: that is, there is reversible 'internal conversion' between low vibrational levels of the upper state and high vibrational levels of the ground state. The larger density of highly excited vibrational states in the ground electronic level will mean that the molecule will spend more of its time in the ground, rather than the excited, electronic condition.† We have hinted several times in this chapter that there is some disagreement about the value of the radiative transition probability, A_f, in NO_2 fluorescence. A_f measured from fluorescence decay observations extrapolated to

†In more complex molecules the density of high vibrational levels is so high in the ground state that the reverse excitation of electronic states by vibrational energy in the ground state is highly improbable.

zero gas pressure is about $2\cdot3 \times 10^4$ s^{-1}, while A_f, calculated via the ν^3 relationship with ν centred on the *emission* maximum from the integrated extinction coefficient over the absorption band, is about 10^6 s^{-1}. At least part of the discrepancy can be understood

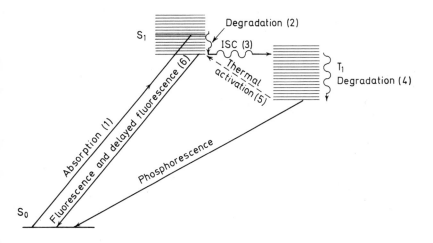

Figure 4.10. Steps in the excitation mechanism for E-type delayed fluorescence

in terms of the reversible energy transfer mechanism: the observed fluorescence decay rate is slower than the 'true' rate because most of the molecules reside in the non-radiating ground electronic state, and the first-order loss rate is not appropriate to the total number of molecules remaining in *some* excited condition.† The process is, therefore, analogous to delayed, thermally activated, fluorescence.

4.7 CHEMILUMINESCENCE

Some chemical reactions are accompanied by the emission of light, and the phenomenon is that of *chemiluminescence*. The excitation

---

†Note that this interpretation means that *true* values of $\pi\sigma_q^2$ for the electronic quenching process need not necessarily exceed $\pi\sigma_{coll}^2$ even if the high value of A_f is chosen, since, on this basis, the measured quenching rates include collisional deactivation of vibrational energy in the ground state.

is *not* thermal; in flames (which show emission characteristic of, for example, the species C_2, CH and OH) emission intensities may be higher than those expected from the flame temperature, and the radiation is chemiluminescent. Several natural chemiluminescent phenomena are well known, among them the light of glow-worms and fireflies, the glow of rotting fish, many bacteria and the cold will o' the wisp.

Detailed understanding of the excitation mechanism is restricted to few systems, mainly involving species in the gas phase. In this section we shall begin by mentioning some processes which are chemiluminescent in solution, and then select some gas-phase chemiluminescent reactions for more detailed discussion.

Chemiluminescent emission may correspond to emission from the reactant or from products. Lucigenin (N, N'-dimethylbiacridylium nitrate) may be oxidised by alkaline H_2O_2 under mild conditions to give a green chemiluminescence which is identical with fluorescent emission from lucigenin: the reactant is acting as a catalyst for the decomposition of H_2O_2, although the mechanism of electronic excitation is not certain. In hot solutions lucigenin is oxidised to N-methyl acridone, and blue emission is observed, corresponding to the fluorescence of this product. The oxidation of luminol (3-amino-cyclicphthalhydrazide) in alkaline solution by ferricyanide,

$$+ \quad O_2 \quad \longrightarrow \quad + \quad N_2 \qquad (4.30)$$

is one of the best-known chemiluminescent reactions. The green-blue emission derives from the aminophthalate ion. Many oxidation reactions which involve organic peroxides or hydroperoxides emit a narrow band at $\lambda = 634$ nm; the same band is emitted from the reaction of sodium hypochlorite and H_2O_2. This band has been shown to be a 'dimol' emission involving *two* excited oxygen molecules in the $^1\Delta_g$ state. Radiation of the $^1\Delta_g \rightarrow {}^3\Sigma_g^-$ system is highly forbidden for electric dipole interactions, but emission is observed weakly, as a result of a magnetic dipole transition, at

$\lambda = 1269$ nm (near infra-red). The band at $\lambda = 634$ nm is equivalent to twice the energy of $O_2(^1\Delta_g)$ and arises when a collision complex $O_2(^1\Delta_g):O_2(^1\Delta_g)$ loses the whole of its excitation in a single quantum transition. If fluorescent substances (e.g. 9:10-substituted anthracenes) are present in the oxidation systems, then the characteristic fluorescence of these substances may be excited by intermolecular energy transfer (Chapter 5) from the energy-rich products of reaction.

A most remarkable chemiluminescent substance has recently been developed for possible commercial applications. Tetrakis-(dimethylamino) ethylene oxidises spontaneously in air to produce a very bright and long-duration green chemiluminescence. The oxidation mechanism is very complex. Emission appears to be from excited molecules of the parent substance: how they obtain their energy is not entirely clear. The material has been proposed for use, incorporated in a rope, as a safety light (e.g. for night-time traffic accidents) which will become illuminated when an evacuated covering is torn off.

Two main types of process lead to chemiluminescence in the gas phase: recombination reactions and exchange reactions. Although chemiluminescent two-body recombination reactions are known, three-body recombination,

$$A + B + M \xrightarrow{k_c} AB^* + M \tag{4.31}$$

(in which the third body M stabilises the newly formed species AB; A + B is usually atom + atom, or atom + small molecule) is a more common source of intense chemiluminescence. If emission and quenching processes,

$$AB^* \xrightarrow{A_c} AB + h\nu \tag{4.32}$$

$$AB^* + M \xrightarrow{k_q} AB + M \tag{4.33}$$

are the only significant fates of AB*, then the ordinary steady state treatment yields a value for the intensity of chemiluminescence, I_c,

$$I_c = \frac{A_c k_c [A] [B] [M]}{A_c + k_q [M]} \tag{4.34}$$

if, as is often the case at moderately high pressures, $A_c \ll k_q[M]$, I_c is approximately independent of $[M]$ and the process may *appear* to be second-order. The intensity will, however, depend on the *nature* of M, since both k_c and k_q vary with the chemical species.

Many of the investigations of gas-phase chemiluminescence of atom recombination reactions have been performed in flow systems using an electric discharge to produce relatively high concentrations of atoms (e.g. 1%–10%). The atomic species are usually in their ground electronic state, and it is therefore not surprising that three-body chemiluminescence normally shows emission from levels *just* below the dissociation threshold for normal (unexcited) fragments. However, it is often found that the emission originates from an electronic state that does *not* correlate (i.e. lie on the same potential curve or surface) with ground state A and B; it seems that the emitting state is populated by radiationless transitions from a state which *does* correlate with the normal species. Thus reaction (4.31) is a gross oversimplification of the excitation process, and several detailed mechanisms can be visualised which give overall third-order kinetic behaviour. Since one of the states with which normal A and B correlate may be the ground state of AB, a considerable fraction of AB might be formed unexcited and thus not give rise to chemiluminescence:

$$A + B + M \xrightarrow{\ k'_c\ } AB + M \tag{4.31a}$$

One of the surprises about gas-phase chemiluminescence is the relatively large fraction of newly formed molecules that do, in fact, emit.

The familiar yellow afterglow of 'active' nitrogen (nitrogen that has been subjected to an electric discharge) is a result of chemiluminescent recombination of ground state (^{4}S) nitrogen atoms. Most of the visible emission is of the 'First Positive' band system ($^3\Pi_g \rightarrow {}^3\Sigma_u^+$), so that reaction (4.31) must be written

$$N(^4S) + N(^4S) + M \longrightarrow N_2(^3\Pi_g) + M \tag{4.35}$$

even though two ^{4}S atoms do not correlate with the $^3\Pi_g$ state of N_2. Figure 4.11 shows potential energy curves for some states of N_2. It is thought that the $^3\Pi_g$ state is populated by a (collision-induced) radiationless transition from the $^3\Sigma_u^+$ state. Measurements of the absolute emission efficiency show that between one-third and one-half of the total recombination proceeds via reaction (4.35) to the $^3\Pi_g$ state. If the rate of formation of $^3\Sigma_u^+$ and ground, $^1\Sigma_g^+$, states is in the statistically expected ratio of three to one, the crossing $^3\Sigma_u^+ \rightarrow {}^3\Pi_g$ must be highly efficient (about 66%) at the pressures employed.

The kinetic behaviour of the First Positive band emission is well represented by Eq. (4.34). At pressures above about 1 torr the

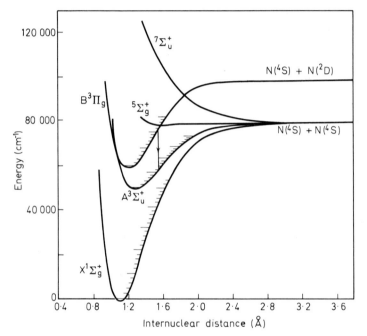

Figure 4.11. Potential energy curves for some states of N_2. (From B. A. Thrush, Chemistry in Britain, **2**, *287 (1966))*

intensity is proportional to $[N]^2$ and independent of $[M]$ but not of its nature; at lower pressures the intensity becomes proportional to $[N]^2[M]$.

Another highly efficient chemiluminescent recombination is the reaction

$$O + NO + M \longrightarrow NO_2^* + M \qquad (4.36)$$

which gives rise to the so-called 'air afterglow'. The intensity is proportional to $[O][NO]$, and is dependent on the nature of the third body, at ordinary pressures; as predicted by Eq. (4.34), the intensity becomes proportional to [M] at sufficiently low pressures (around 5×10^{-2} torr). Bands in the emission spectrum show that the air afterglow involves the same electronic transition of NO_2 as that responsible for the visible absorption; the short wavelength limit of the chemiluminescence corresponds exactly to the onset of predissociation in the absorption spectrum.

Absolute intensity measurements imply that between 50% and 90% of the total recombination produces excited, rather than ground state, NO_2. However, it seems improbable that the rate of recombination into an electronically excited state can be much greater than that into the ground state. An alternative explanation is that electronically excited NO_2 is produced by the 'reverse internal conversion' crossing discussed in the last section in connection with the radiative lifetime of NO_2. The hypothesis calls for the formation of ground state NO_2 with a high degree of vibrational excitation; intramolecular energy transfer then populates lower vibrational levels of the upper electronic state.

Many other recombination reactions are chemiluminescent in the gas phase. The carbon monoxide flame emission bands are a result of $O + CO$ recombination, and the spectrum shows features due to transition from the bent $^1B_2$ state of CO_2 to the linear $^1\Sigma_g^+$ ground state. Excitation of atomic spectra in some flames containing metal or metal salt vapours often involves a chemiluminescent recombination of a different kind. The overall excitation process can be represented by

$$H + X + M \longrightarrow HX + M^* \qquad (4.37)$$

where $X = H$ or OH and M is the metal atom. Other processes in which chemiluminescence has been studied are $N + O$, $O + O$, $H + NO$, $O + SO$, and various halogen atom recombinations.

The photochemistry we have discussed so far in this book has always involved electronically excited states. A number of exothermic transfer reactions of the general type

$$A + BC \longrightarrow AB^v + C \qquad (4.38)$$

yield a product molecule AB with a high degree of vibrational excitation in the newly formed bond (the superscript v is used to represent this excitation). This vibrational excitation does not correspond to the kinetic temperature of the gas, so that, if the vibration is 'infra-red active' (i.e. has associated with it an oscillating dipole), chemiluminescent emission may be observed in the red or near-infra-red regions of the spectrum. The reaction of atomic hydrogen with halogens is typical of the reactions which lead to infra-red chemiluminescence,

$$H + X_2 \longrightarrow HX^v + X \qquad (4.39)$$

With $X_2 = Cl_2$, vibration–rotation emission spectra for transitions of $\Delta v = 0$, 1 and 2 have been observed. At pressures low enough to avoid collisional degradation, up to six vibrational quanta are present in the newly formed HCl. The vibrational energy bears no relation to an equilibrium distribution, although the fine structure of the spectrum suggests that rotational energy is in thermal equilibrium at a temperature not more than 100°C above that of the reaction vessel. The exothermicity of the reaction is rather less than the energy of the observed six vibrational quanta; however, the total energy available for excitation is that from the *top of the activation barrier* to the product HCl (i.e. equivalent to $E_a + |\Delta H|$).

Vibrational chemiluminescence from exchange reactions currently arouses great interest because of its applications in studies of theoretical kinetics. Computer studies of reaction trajectories on hypothetical potential surfaces linking products and reactants can reveal the expected vibrational and rotational excitation in the products. The detailed spectroscopic structure of vibrational chemiluminescence can be used for comparison with the computer-calculated excitation, and parameters for empirical or semi-empirical reaction surfaces can be adjusted for the best fit of calculated and experimental data. The technique has proved most successful in deriving a surface which satisfactorily describes the H + halogen reaction. The complexities of the calculations have, however, so far restricted its use to three-atom (A + BC) reactions, although several four-atom

$$A + BCD \longrightarrow AB^v + CD \qquad (4.40)$$

reactions display infra-red chemiluminescence. An example of such a process is the highly exothermic (80 kcal (335kJ) mol^{-1}) reaction

$$H + O_3 \longrightarrow OH^v + O_2 \qquad (4.41)$$

The observed OH emission extends from the infra-red into the long wavelength end of the visible spectrum, and the bands are forbidden 'overtone' vibrational transitions ($\Delta v = 4$ or 5). The overtone bands of OH are observed in the glow of the night sky, and reaction (4.41) is believed to be the source of OH excitation in the upper atmosphere.

Electronic excitation also occurs occasionally in transfer reactions. For example, it is thought that the exothermic process

$$CH_3O + OH \longrightarrow CH_2O^* + H_2O + 101 \text{ kcal} (423kJ) \text{ mol}^{-1} \quad (4.42)$$

accounts for the appearance of formaldehyde emission bands in the *cool flames* observed in the oxidation of hydrocarbons. We shall select for comparison and more detailed discussion the two chemiluminescent reactions

$$NO + O_3 \longrightarrow NO_2^* + O_2 + 49 \text{ kcal} (205kJ) \text{ mol}^{-1} \quad (4.43)$$

and

$$SO + O_3 \longrightarrow SO_2^* + O_2 + 106 \text{ kcal} (444kJ) \text{ mol}^{-1} \quad (4.44)$$

The short wavelength limit of the $NO + O_3$ chemiluminescence corresponds to an energy of $48 \cdot 4$ kcal $(203kJ)$ mol^{-1}; that of the $SO + O_3$ emission is equivalent to 101 kcal $(423kJ)$ mol^{-1}. Since these energies are only slightly smaller than the exothermicities of the processes, there is little doubt that the lower states of both transitions are the ground states of the product NO_2 or SO_2. Emission in the $NO + O_3$ reaction consists of transitions from the $^2B_1$ state to the ground, $^2A_1$, state. The ground state of SO_2 is $^1A_1$, and both singlet, ^1B, and triplet, ^3B, excited states emit to it in the $SO + O_3$ chemiluminescence. Measurements of absolute emission intensities and of overall rates of reaction, combined with fluorescence quenching data for the excited states, have allowed calculation of pre-exponential factors and activation energies for reactions (4.43), (4.44) and the equivalent reactions to unexcited products. The results are given in Table 4.7.

Table 4.7. Rate data for reaction to ground state products and for excitation in the reaction of NO and SO with O_3.[a] Data from B. A. Thrush, *Chemistry in Britain* **2**, 287 (1966)

	$NO + O_3$ *reaction*				$SO + O_3$ *reaction*		
Product	ΔH (kcal mol^{-1})	E_a	A (l mol^{-1}s^{-1})	Product	ΔH (kcal mol^{-1})	E_a	A (l mol^{-1}s^{-1})
Ground ($^2A_1$)	-48	$2 \cdot 4$	5×10^8	Ground ($^1A_1$)	-106	$2 \cdot 1$	$1 \cdot 5 \times 10^9$
Excited ($^2B_1$)	~ -9	$4 \cdot 2$	5×10^8	Excited (^1B)	-21	$4 \cdot 2$	10^{8b}

[a]Original source gives energies in calorie units: 1 cal = $4 \cdot 184$ J.
[b]It is not possible to give an absolute value for **A** for the formation of the triplet state of SO_2, since fluorescence studies indicate that much of the observed triplet emission is due to quenching from the ^1B state. However, E_a for the emission ($3 \cdot 9$ kcal $(16kJ)$ mol^{-1}) is somewhat lower than E_a for the singlet emission, and at least some $SO_2(^3B)$ is populated directly.

For both reactions, the activation energies for emission are higher than those for the more exothermic reaction to form ground state products. In the $NO + O_3$ reaction the whole of the difference in rate of formation of ground and excited species is a result of the higher activation energy for the latter process, since the pre-exponential factors are virtually identical. In contrast, the pre-exponential factor for the $SO + O_3$ reaction is more than an order of magnitude smaller for the excitation process than for the overall reaction. This difference between the two reactions probably arises from differences in the potential energy surfaces available for the reaction. The degeneracy associated with the orbital momentum of NO (a Π state) makes available two potential energy surfaces for the reaction: these surfaces have a spin multiplicity of two $(S = \frac{1}{2})$, since they derive from $NO(S = \frac{1}{2})$ and $O_3(S = 0)$. Let us consider the electronic configuration in NO_2 (17 valence electrons). CO_2, possessing 16 valence electrons, is linear; if NO_2 were linear, the additional electron would occupy one of a degenerate pair of π orbitals, thus yielding a $^2\Pi$ state. The degeneracy is removed by bending the molecule, there being a decrease in the energy of the state with the electron in the π orbital lying in the plane of the molecule: this state is the ground, $^2A_1$, state. Promotion of the electron to the perpendicular π orbital gives the nearly linear $^2B_1$ excited state. We can now see why the pre-exponential factors for formation of $NO_2(^2A_1)$ or $NO_2(^2B_1)$ are similar. As NO and O_3 approach, they do so along a surface which correlates initially with linear $NO_2(^2\Pi)$, the ground or excited state of NO_2 being formed by bending of this transition state. Thus the frequency factors for the two processes are similar, and the two activation energies arise from the two possible orientations of the π orbital with respect to the O—N . . . O plane in the transition state.

For the $SO + O_3$ reaction, involving orbitally non-degenerate species, only one reaction surface (a triplet, since SO is $^3\Sigma^-$) is available, and it is clear that this correlates with ground state products, $SO_2(^1A_1) + O_2(^3\Sigma_g^-)$. Electronically excited $SO_2 +$ ground state O_2 must correlate with *other* triplet surfaces. The appearance of $SO_2(^1B)$ or $SO_2(^3B)$ requires, therefore, spin-permitted intramolecular energy transfer from one surface to another. It is not to be expected that crossing should occur for all molecules, and it is not surprising that the frequency factor (**A**) for the process involving crossing should be much less than that for the process forming $SO_2(^1A_1)$.

We have seen in this section that species excited in chemical reactions take part in exactly the same processes (emission, quenching, intramolecular energy transfer) as those formed by absorption of light. It is hoped that the discussion of chemiluminescent phenomena has also shown the interrelation between reaction kinetics, spectroscopy and photochemistry.

BIBLIOGRAPHY

E. J. BOWEN (Ed.), *Luminescence in chemistry,* Van Nostrand, London (1968)

E. J. BOWEN, 'Light emission from organic molecules', *Chemistry in Britain* **2,** 249 (1966)

C. A. PARKER, 'Photoluminescence as an analytical technique', *Chemistry in Britain* **2,** 160 (1966)

C. A. PARKER, 'Phosphorescence and delayed fluorescence from solutions', *Adv. Photochem.* **2,** 305 (1964)

B. A. THRUSH, 'Formation of electronically excited molecules in simple gas reactions', *Chemistry in Britain* **2,** 287 (1966)

J. C. POLANYI, 'Energy distribution among reaction products and infrared chemiluminescence', *Chemistry in Britain* **2,** 151 (1966)

C. A. PARKER, *Photoluminescence of solutions,* Elsevier, Amsterdam (1969)

5

Energy transfer: Emission processes (2)

5.1 INTERMOLECULAR ENERGY TRANSFER

The *intra*molecular exchange of electronic energy between different states, and its exchange with vibrational energy, play an important part in determining the photochemical behaviour of a molecule (Chapter 4). *Inter*molecular exchange of energy between two discrete partners may also take place; the 'acceptor' (A), which receives excitation from the 'donor' (D), then participates in those processes open to it as an electronically excited species. *Photosensitised* phenomena, in which the change of interest occurs in a species other than the one that absorbed radiation, are believed to be of great significance in photobiology; they also provide valuable insights into photophysical processes.

Franck predicted in 1922 that electronic excitation could be exchanged between atoms, and Cario and Franck subsequently demonstrated *sensitised fluorescence* in a mixture of mercury and thallium vapours. The mixture was irradiated with the $\lambda = 253\cdot7$ nm resonance line of mercury, to which thallium vapour is transparent; emission was observed from the thallium. Absorption of light by Hg raises it to the resonance level, $^3P_1$, and energy is then transferred to the thallium:

$$Hg + h\nu_{\lambda = 253\cdot7 \text{ nm}} \longrightarrow Hg^* \tag{5.1}$$

$$Hg^* + Tl \longrightarrow Hg + Tl^* \tag{5.2}$$

$$Tl^* \longrightarrow Tl + h\nu \tag{5.3}$$

Energy transfer in organic fluorescence was recognised by the increase with concentration in depolarisation of emission excited by polarised light in dyes. At the higher concentrations energy is

129

exchanged between dye molecules before emission occurs, and the emitting molecule is not the same as the absorbing one.

Exchange of energy between two different species is not so restricted with regard to exact equivalence of internal energy between initial and final states as in the case of intramolecular exchange, since an energy excess can be taken up by translation (or, more rarely, a deficiency supplied by the kinetic energy of collision). Ten different types of energy exchange can be classified according to the modes (electronic, vibration, rotation and translation) between which the exchange occurs; except in the rare case of exact energy resonance, some energy is always converted to or from translation. We have already given implicit consideration to the degradation of electronic excitation to vibration, rotation or translation, in our discussion of physical quenching of fluorescence. Direct demonstration that electronic energy may be converted to vibrational excitation is afforded by the process

$$Hg^*(^3P_1) + CO(v = 0) \longrightarrow Hg(^1S_0) + CO^v(v \leqslant 8) \qquad (5.4)$$
$$\text{vibrationally}$$

Infra-red-sensitised (vibrational) fluorescence of hot CO is observed. (We should note that, unless otherwise stated, $Hg^*(^3P_1)$ is always created by absorption in mercury vapour of the resonance line at $\lambda = 253 \cdot 7$ nm.) The reverse process, vibrational to electronic exchange, also occurs; shock waves in N_2 or CO can produce a high degree of vibrational excitation which has been shown to excite the yellow emission D lines of sodium:

$$N_2^v(v > 0) + Na(^2S) \longrightarrow N_2(v = 0) + Na^*(^2P) \qquad (5.5)$$

Much of the following discussion will, however, be directed towards electronic–electronic energy exchange, and it will be assumed that *excess* energy goes into other modes of excitation. Where the absorption spectrum of the acceptor overlaps the emission spectrum of the donor, there are, of course, quantised vibronic levels of A and D for which energy exchange is isoenergetic, and no increase in kinetic energy is needed. For electronic energy exchange between atoms, some translational energy is almost always released.

Several different mechanisms of electronic energy transfer are believed to operate under different circumstances. The first of these is the so-called 'trivial' mechanism of radiative transfer, which can be represented by the processes

$$D^* \longrightarrow D + h\nu \tag{5.6}$$

$$A + h\nu \longrightarrow A^* \tag{5.7}$$

The mechanism is trivial in name and simplicity only, since it is the one energy transfer mechanism that can operate over very large separation of D and A: the interaction necessarily follows the laws of light propagation. Radiative energy transfer is all-important to our existence, because it is how we receive the energy of reactions occurring in the sun; and the related radiative energy transfer processes occurring in upper and lower atmospheres establish the temperature equilibria and meteorological conditions upon which we depend. The efficiency of radiative transfer is a function of the overlap between the emission spectrum of D and the absorption spectrum of A (a factor which appears in all transfer mechanisms), and also of the size and shape of the sample: since D* will emit in all directions, the probability of radiative transfer increases with sample volume. It will be obvious that experiments designed to study non-radiative energy transfer must eliminate or make due allowance for the radiative process.

'Short-range' energy transfer arising from *exchange interaction* occurs over intermolecular or interatomic distances (henceforth referred to as r) not much exceeding the collision diameter; the interaction decreases in a complex fashion with r raised to a high power.

'Long-range' transfer may arise by sequential short-range excitation of many species so that the excitation appears ultimately at a place 'distant' from the original location of excitation. Long-range transfer is, however, predicted to occur also by a direct mechanism involving electrical, or Coulombic, interactions between transition dipoles (or higher multipoles). These multipoles are those involved in optical interactions with the electric vector of radiation; the usual optical selection rules apply to both the transitions $D^* \to D$ and $A \to A^*$, and dipole–dipole (dd) interactions are stronger than dipole–quadrupole (dq) interactions, and so on. For a dd interaction, theory predicts that the strength of the interaction should fall off as $1/r^6$, and relatively long-range energy exchange becomes possible.

An even longer-range transfer, showing a $1/r^3$ dependence, may occur in crystals, solid solutions and some fluids, as a result of *exciton* migration. The concept of the exciton was introduced by

Frenckel to interpret certain crystal spectra; an electron–hole pair was looked upon as an entity which could move about the crystal as a result of interactions between lattice sites. For our purposes, we can regard the electronic excitation in an irradiated species as an exciton which is free to wander over a considerable number of lattice sites.

Energy may be transferred from one excited species, D*, to another, A*, already possessing some excitation, to raise the latter to a higher electronic state A**,

$$D^* + A^* \rightarrow D + A^{**} \tag{5.8}$$

This process of *energy-pooling* must be relatively slow at low concentrations of D* and A*, since bimolecular collision of the excited species will be a rare event. Formation of excited species by energy-pooling has, however, been recognised in several systems, and provides the mechanism for *P-type delayed fluorescence*. Energy-pooling may also permit the occurrence of chemical reactions which require more energy than that available in a single quantum of radiation, and such energy storage may be a necessary step in several photobiological systems.

5.2   SHORT-RANGE, COLLISIONAL, ENERGY TRANSFER

Energy transfer by exchange interaction may be thought of as a special kind of chemical reaction in which the chemical identity of the partners A and D does not change, but excitation is transferred from one to the other. The transition state is then expected to possess a separation between A and D not greatly different from the sum of the gas-kinetic collision radii, and energy transfer by the exchange mechanism is probably only for values of r of this order. In common with other chemical processes, energy transfer can be efficient only if the potential energies of reactants and products are connected by a *continuous* surface which describes the potential energy of the system as a function of the several interatomic distances; a reaction occurring 'on' such a surface is said to proceed *adiabatically*. In other words, the reactants and products must correlate with each other and with the transition state. Most chemical reactions involving ground state partners can occur adiabatically, but in processes such as energy exchange, where several electronic states are involved, the requirement for adiabatic reaction may impose some restrictions on the possible states for A, A* and D, D* if there is to be efficient transfer of excitation. In atoms or small molecules there must be correlation of electron spin, orbital momentum, parity, and so on. However, correlation in complex molecules of low symmetry usually only involves the electron spin. To test for correlation, the possible total spin of the transition state is calculated from the individual spins of the reactants by the addition of the quantum vectors (see Section 2.6 for addition of quantised vectors in single atoms or molecules). Thus, for reactants A and B with spins S_A and S_B, the total spin of the transition state can take magnitudes $|S_A + S_B|$, $|S_A + S_B - 1|, \ldots, |S_A - S_B|$. It is then necessary to see whether the products, X and Y, can also give at least one of the same total spin magnitudes in the transition state. Table 5.1 shows multiplicities in the transition state that can arise from some multiplicities of two separated species. Thus we can see, by use of the table, that all of the processes

Table 5.1. Multiplicities of transition states for given multiplicities of two separated species

Separated species	Transition state
singlet + singlet	singlet
singlet + doublet	doublet
singlet + triplet	triplet
doublet + doublet	singlet, triplet
doublet + triplet	doublet, quartet
triplet + triplet	singlet, triplet, quintet

$$\begin{array}{l} \hookrightarrow [\text{singlet}] \leftrightarrow C(\text{singlet}) + D(\text{singlet}) \\ A(\text{triplet}) + B(\text{triplet}) \leftrightarrow [\text{singlet or triplet}] \leftrightarrow C(\text{doublet}) + D(\text{doublet}) \\ \rightsquigarrow [\text{singlet or triplet}] \leftrightarrow C(\text{triplet}) + D(\text{triplet}) \end{array}$$

$$(5.9)$$

can occur adiabatically. On the other hand, the reaction

$$A(\text{singlet}) + B(\text{triplet}) \nleftrightarrow [\text{no common multiplicity}] \nleftrightarrow C(\text{singlet}) + D(\text{singlet})$$

$$(5.10)$$

cannot proceed (in either direction) adiabatically. Arguments of this kind were used by Wigner to derive the *Wigner Spin Correlation Rules* (e.g. triplet + triplet → triplet + triplet, etc.).

Although a reaction is likely to be efficient only if it is adiabatic, non-adiabatic reactions can also occur.† We may look upon a non-adiabatic reaction as one in which crossing occurs between two intersecting or closely approaching potential energy surfaces. The crossing process is governed by the ordinary selection rules for radiationless transitions. In particular, a spin-forbidden reaction cannot proceed adiabatically because no common spin states can be written for the transition complex, and potential surfaces for transition states derived from reactants and products must be of different multiplicity. Hence, $\Delta S \neq 0$ for the *intramolecular* energy transfer, the crossing is of low probability (cf. Section 4.5), and the efficiency of the non-adiabatic *intermolecular* energy transfer is small.

Franck suggested in his first predictions of sensitised fluorescence that energy transfer should be most efficient when the energy gap (ΔE) is small; in general, this expectation has been borne out by experimental results. Energy transfer between atoms is particularly likely to show a clear relationship between transfer efficiency and ΔE, since there are no internal degrees of freedom for accommodating either positive or negative energy differences. The efficiency of energy transfer between atoms is especially likely to be small if $\Delta E > 0$ (i.e. excitation energy of acceptor greater than energy of excited donor), since the excess energy will have to be supplied as kinetic energy of collision. The activation energy for the transfer will be at least ΔE, and the rate of reaction will therefore be reduced below the collision rate by a corresponding factor (cf. Table 4.7 for a similar situation in intramolecular energy transfer).

As an example of the inverse dependence on ΔE of efficiency of energy transfer, we may consider mercury-sensitised emission from sodium vapour. Irradiation by the Hg-resonance line at $\lambda = 253 \cdot 7$ nm populates $Hg^*(^3P_1)$, and both transfer processes

$$Hg^*(6^3P_1) + Na(3^2S) \rightarrow Hg(6^1S_0) + Na^*(9^2S) \quad \Delta E = +162 \, cm^{-1} \tag{5.11}$$

$$Hg^*(6^3P_1) + Na(3^2S) \rightarrow Hg(6^1S_0) + Na^*(4^2D) \quad \Delta E = -4839 \, cm^{-1} \tag{5.12}$$

---

†The following discussion is an oversimplification, but adequate for the present purposes. Reactions which are by common consent called non-adiabatic—such as those which are spin-forbidden—may in fact occur adiabatically but with low efficiency.

(quoted energies in the case of non-degenerate doublets refer to an approximate average value) can be shown to occur, since emission of the $Na(9^2S \rightarrow 3^2P)$ and $Na(4^2D \rightarrow 3^2P)$ doublets is observed. The intensity of the sensitised $4^2D \rightarrow 3^2P$ lines is less than that of the $9^2S \rightarrow 3^2P$ lines, although in normal emission from excited sodium the relative intensity of the emission from the 4^2D is greater than that from the 9^2S state. In the sensitised fluorescence experiment, it would appear that $Na(9^2S)$ is populated preferentially because of the near-resonance ($\Delta E = +162 \text{ cm}^{-1}$) of the energy transfer process. Further confirmation that energy resonance is important to transfer comes from the effect of added nitrogen on the Hg-sensitised fluorescence of sodium: there is a substantial increase in the relative intensity of the $Na(7^2S \rightarrow 3^2P)$ lines. Nitrogen is known to quench $Hg(6^3P_1)$ to metastable $Hg(6^3P_0)$,

$$Hg^*(6^3P_1) + N_2 \longrightarrow Hg^*(6^3P_0) + N_2 \qquad (5.13)$$

This state can excite Na to its 7^2S state with only a small ΔE,

$$Hg^*(6^3P_0) + Na(3^2S) \longrightarrow Hg(6^1S_0) + Na^*(7^2S) \; \Delta E = 565 \text{ cm}^{-1}$$

$$(5.14)$$

An exception to the general rule appeared to be found in the system first studied: Hg-sensitised fluorescence of thallium. At 900°C the process

$$Hg^*(6^3P_1) + Tl(5^2P) \longrightarrow Hg(6^1S_0) + Tl^*(6^2D) \; \Delta E \sim -3250 \text{ cm}^{-1}$$

$$(5.15)$$

is apparently *more* efficient than the process

$$Hg^*(6^3P_1) + Tl(5^2P) \longrightarrow Hg(6^1S_0) + Tl^*(8^2S) \; \Delta E = -666 \text{ cm}^{-1}$$

$$(5.16)$$

It now appears, however, that species other than $Hg^*(6^3P_1)$ may contribute to the excitation of $Tl^*(6^2D)$.

Energy exchange in molecules probably also occurs so as to minimise the amount of kinetic (translational) energy that must be liberated; thus we may expect exchange of energy between vibronic levels in near resonance. Excess vibrational energy will be degraded rapidly (at least in condensed-phase systems), and the acceptor molecule left in its ground vibrational state. The *apparent* energy gap is therefore the electronic difference between D* and A*, although this is *not* ΔE for the actual transfer process. An anomaly in the dependence on ΔE of rate of transfer arises because of the difference between actual and apparent energy gaps. Table 5.2 shows measured rate constants for several triplet–triplet energy exchange processes

$$D^*(T_1) + A(S_0) \longrightarrow D(S_0) + A^*(T_1) \qquad (5.17)$$

and lists the energy difference between $D^*(T_1, v = 0)$ and $A^*(T_1, v = 0)$; in each case D is biacetyl.

Table 5.2. Rate constants for triplet energy exchange from biacetyl in benzene solution. Data of K. Sandros and H. L. J. Backström, *Acta. chem. scand.* **12**, 828 (1958)

Acceptor A	$\Delta E(\text{cm}^{-1})$	Rate constant for transfer $(\text{l mol}^{-1}\text{s}^{-1})$
3, 4-Benzpyrene	5000	8.2×10^9
Anthracene	5000	8.1×10^9
1, 2-Benzanthracene	3200	7×10^9
Pyrene	3000	7.5×10^9
trans-Stilbene	2000	4.4×10^9
Coronene	700	2×10^8
1-Nitronaphthalene	500	1.1×10^8
2, 2'-Binaphthyl	150	9.7×10^6

Contrary to first expectations, the rate constant is highest for the largest value of ΔE. The rate constants refer, however, to the *net* rate of formation of A*, and allow for the re-excitation of biacetyl by the reverse of reaction (5.17). Since vibrational degradation is effectively complete, this re-excitation process has an activation energy of about ΔE, and the difference between forward and reverse rates is therefore larger the greater ΔE (but see also p. 140). Reverse excitation also occurs in the gas phase, its importance being determined by the value of ΔE for the reverse process, and by the relative concentrations of acceptor and donor. In solution, re-excitation may proceed more rapidly than the concentration of donor would lead one to suppose, since many collisions between the transferring partners may occur during a single encounter.

The efficiency of energy transfer occurring by the exchange interaction mechanism is related to whether the process can take place adiabatically, but *not* to whether optical selection rules permit radiative transitions in donor and acceptor; this behaviour is one way in which exchange interaction and long-range Coulombic interaction may be distinguished. For example, in the exchange interaction excitation of triplet states by triplet benzophenone, the efficiency of energy transfer is roughly the same both for naphthalene and for 1-iodonaphthalene. We saw in the last chapter (p. 115) that the $T_1 \rightarrow S_0$ radiative transition is by a factor of at least 1000 more probable in the substituted molecule, so that in this case the optical transition probability in the naphthalene molecule does not seem to affect the probability of energy transfer to it. Table 5.3 shows the relative efficiencies of some pairs of processes involving energy transfer from $Hg(6^3P_1)$ to various metal atoms. For each

Table 5.3. Relative efficiencies of pairs of excitation reactions. Data from A. B. Callear in *Photochemistry and reaction kinetics* (Ed. P. G. Ashmore, F. S. Dainton and T. M. Sugden, Cambridge University Press (1967))

Acceptor	Electronic States initial	final	Optical transition	ΔE (cm^{-1})	Preferred reaction
Na	3^2S	9^2S	F	$+162$	preferred
	3^2S	8^2P	A	-113	—
Na	3^2S	7^2D	F	-212	preferred
	3^2S	7^2P	A	-871	—
In	5^2P	7^2P	F	-440	⎱ equally
	5^2P	6^2D	A	-314	⎰ probable

reaction the donor transition, $Hg(6^3P_1 \rightarrow 6^1S_0)$, is formally spin-forbidden, although only mildly so. There is no apparent correlation between transfer probability and forbiddenness in the acceptor transition (if anything, forbidden optical transitions are favourable for energy exchange!). On the other hand, adiabatic reaction can occur in every case. Each reaction involves the adiabatic spin process

$$\text{triplet} + \text{doublet} \longrightarrow \text{singlet} + \text{doublet} \qquad (5.18)$$

and the orbital momentum changes can also take place adiabatically. The values of ΔE are all 'small' (except for the excitation of $Na7^2P$, which is a relatively unfavoured reaction), and probably do not determine the excitation efficiency.

Energy transfer and emission from an excited donor are competitive, and kinetic analysis of sensitised emission phenomena can yield information about rate constants for the transfer process.

A simple excitation scheme (in which A does not absorb) can be written for sensitised emission:

		rate	
$D + h\nu \to D^*$	absorption	I_{abs}	(5.19)
$D^* \to D + h\nu_D$	donor emission	$A_D[D^*]$	(5.20)
$D^*[+M] \to D[+M]$	donor energy loss	$k_D^q[D^*]$	(5.21)
$D^* + A \to D + A^*$	energy transfer	$k_e[A][D^*]$	(5.22)
$A^* \to A + h\nu_A$	acceptor emission	$A_A[A^*]$	(5.23)
$A^*[+M] \to A[+M]$	acceptor energy loss	$k_A^q[A^*]$.	(5.24)

Reactions (5.21) and (5.24) include all non-radiative loss processes

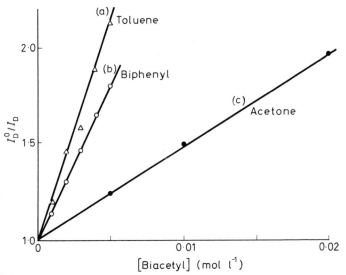

*Figure 5.1. Plot of I_D^0/I_D against biacetyl concentration for quenching by biacetyl of fluorescence from (a) toluene, (b) biphenyl and (c) acetone. (Data of F. Wilkinson and J. T. Dubois, J. chem. Phys. **39**, 377 (1963))*

for D^* and A^*, respectively (e.g. radiationless intramolecular energy degradation and bimolecular quenching). For constant $[M]$ these processes may be described by a single pseudo-first-order rate constant k^q. Solution of the stationary state equations for $[D^*]$ and $[A^*]$ yields the results

$$I_D = \frac{A_D I_{abs}}{A_D + k_D^q + k_e[A]} \tag{5.25}$$

and

$$I_A = \frac{A_A k_e[A] I_{abs}}{(A_A + k_A^q)(A_D + k_D^q + k_e[A])} = \frac{A_A}{A_D} \cdot \frac{k_e[A]}{A_A + k_A^q} \cdot I_D \tag{5.26}$$

where I_D, I_A are the intensities of emission from donor and acceptor, respectively. Thus, if I_D^0 is the emission intensity from the donor when $[A] = 0$, then

$$\frac{I_D^0}{I_D} = 1 + \frac{k_e}{A_D + k_D^q}[A] \tag{5.27}$$

That is, the quenching by the acceptor of donor emission follows a Stern–Volmer law, and values for k_e may be calculated if $(A_D + k_D^q)$ is known. Figure 5.1 shows some results for the quenching by biacetyl of fluorescence from various donors. From the slope of the graph and the known value of $(A_D + k_D^q)$ it may be shown, for example, that $k_e = 3\cdot7 \times 10^{10} \, \mathrm{l \, mol^{-1} \, s^{-1}}$. For transfer from toluene a further check on k_e is available, via Eq. (5.26), from the ratio of acceptor to donor intensities, I_A/I_D, if $(A_A + k_A^q)$ is known. If, as may be the case, A also absorbs the exciting radiation, Eq. (5.26) must be modified; Eqs. (5.25) and (5.27) remain unaltered. Table 5.4 shows rate constants for singlet–singlet energy transfer from several donors to biacetyl in hexane solution.

Table 5.4. Rate constants for energy transfer to biacetyl (hexane solution at 28°C). Data of F. Wilkinson and J. T. Dubois, *J. chem. Phys.* **39**, 377 (1963); and J. T. Dubois and R. L. Van Hemert, *J. chem. Phys.* **40**, 923 (1964)

Donor	$k_e(\mathrm{l \, mol^{-1} s^{-1}})$
Benzene	$3\cdot3 \times 10^{10}$
Toluene	$3\cdot7 \times 10^{10}$
o-Xylene	$3\cdot5 \times 10^{10}$
m-Xylene	$3\cdot3 \times 10^{10}$
p-Xylene	$3\cdot4 \times 10^{10}$
Pentamethylbenzene	$4\cdot5 \times 10^{10}$
Hexamethylbenzene	$4\cdot0 \times 10^{10}$
Ethylbenzene	$3\cdot6 \times 10^{10}$
n-Propylbenzene	$4\cdot2 \times 10^{10}$
n-Butylbenzene	$3\cdot4 \times 10^{10}$
Naphthalene	$2\cdot2 \times 10^{10}$

The large rate constants and their relative insensitivity to the nature of the donor together suggest that the transfer of energy is diffusion-controlled. The Debye equation for the rate constant in a diffusion-controlled reaction of species of similar size (Eq. 4.8) gives a value of $k_e \sim 2.4 \times 10^{10} \, l \, mol^{-1} \, s^{-1}$ for hexane at 28°C, in qualitative agreement with the data of Table 5.4. Even better agreement is obtained if the equation for the diffusion-controlled rate constant is modified for the case where there is no friction between partners: the corrected expression is

$$k_e \sim \frac{8RT}{2\eta} \times 10^3 \, (l \, mol^{-1} \, s^{-1}) \qquad (5.28)$$

which gives $k_e \sim 3.5 \times 10^{10} \, l \, mol^{-1} \, s^{-1}$ in hexane (28°C). The rate constants for *triplet* energy transfer *from* biacetyl to various acceptors, shown in Table 5.2, also approach the diffusion-controlled value when ΔE is large. Note that the relative lack of dependence of k_e on ΔE in the singlet–singlet exchange (Table 5.4) may be related mainly to the much larger values of ΔE in the singlet–singlet processes listed (e.g. perhaps as much as $16\,000 \, cm^{-1}$ for benzene–biacetyl) compared with those for the triplet energy transfer (the largest in Table 5.2 is $5000 \, cm^{-1}$).

Triplet–triplet energy transfer is, in fact, sometimes treated as though it were a different phenomenon from singlet–singlet transfer. However, so far as the exchange interaction mechanism is concerned, the fact that both A and D change their spin multiplicity is of no account, since the reaction is adiabatic. Observed differences in photochemical behaviour arise from the long radiative lifetimes of triplet states. In media in which quenching and radiationless energy degradation are slow (e.g. in rigid glasses) the long actual lifetime of the triplet donor means that even inefficient energy transfer can compete successfully with other loss processes. At the same time, sensitised phosphorescence is likely to be seen only in systems where radiationless loss and quenching are not the major fates of the triplet acceptor (e.g. again in rigid glasses, or with acceptors such as biacetyl).

Absorption of the exciting radiation by both donor and acceptor can complicate interpretation of sensitised fluorescence in studies of singlet–singlet transfer. Triplet–triplet exchange, on the other hand, can be investigated in systems where only the donor absorbs. Suitable choice of donor and acceptor permits the triplet of D to lie *above* that of A, so that $D^* \rightarrow A$ transfer can occur, at the same time that $S_1(D)$ is below $S_1(A)$, so that excitation of $S_1(D)$ can occur at wavelengths longer than those absorbed by A. The required order of energy levels is often found for aromatic carbonyl compounds as donors, and aromatic hydrocarbons as acceptors. Figure 5.2 shows the energies of the triplet and singlet states in benzophenone and naphthalene. Irradiation at $\lambda = 366$ nm of a benzophenone–naphthalene mixture (in a rigid glass at $-180°$C) leads

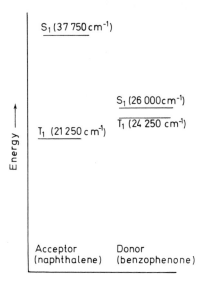

Figure 5.2. Energy levels of excited singlet and triplet states in naphthalene and in benzophenone. (Data from V. L. Ermolaev, Soviet Phys., Usp. 333 (Nov.-Dec., 1963))

to emission from naphthalene, and the spectrum is identical with that of naphthalene phosphorescence. Light at $\lambda = 366$ nm is, however, absorbed only by the benzophenone, and the emission from naphthalene is *sensitised* phosphorescence. The excitation scheme is

$$B(S_0) + h\nu_{\lambda \sim 366\,nm} \rightarrow B(S_1) \tag{5.29}$$

$$B(S_1) \overset{ISC}{\rightsquigarrow} B(T_1) \tag{5.30}$$

$$B(T_1) + N(S_0) \rightarrow B(S_0) + N(T_1) \tag{5.31}$$

$$N(T_1) \rightarrow N(S_0) + h\nu_{phos} \tag{5.32}$$

Direct identification by ESR of the triplet state of naphthalene has proved possible in similar glasses of benzophenone–naphthalene mixtures irradiated at $\lambda = 366$ nm.

The absolute quantum yield for the sensitised phosphorescence of naphthalene is 0·07. For naphthalene on its own, irradiated in the first $S_1 \leftarrow S_0$ absorption band, $\phi_p = 0·03$, so that the sensitised process is more efficient than the unsensitised one. Intersystem crossing $(S_1 \rightsquigarrow T_1)$ is more probable in benzophenone than in naphthalene ($\phi_{ISC} = 0·99$ for benzophenone, $= 0·39$ for naphthalene, *at 29°C in benzene*), and the intermolecular triplet–triplet energy transfer clearly proceeds so efficiently that the triplet of naphthalene is populated more readily via the intermolecular process than it is by intramolecular ISC from its own S_1 level.

Much of our discussion about energy transfer processes has suggested that population of acceptor levels *above* those of the donor (i.e. with ΔE positive) can occur only with an activation energy for reaction equal to ΔE. Indeed, triplet–triplet energy transfer in solution was first demonstrated, albeit circumstantially, by the quenching of biacetyl phosphorescence only by those quenchers whose triplet level lay below that of biacetyl. The implication is that quenching involves triplet–triplet energy exchange, and subsequent experiments have, in fact, detected quencher triplets by their absorption spectra. There are, however, some most interesting cases where, although energy transfer is of reduced efficiency, the activation energy is much smaller than the endothermicity. For example, the rate of transfer to *cis*-stilbene of triplet excitation drops by a factor of rather less than two as the donor energy drops from approximately the same as that of the *cis*-stilbene triplet (~ 57 kcal (238 kJ) mol^{-1}) to 3 kcal (12·6 kJ) mol^{-1} less. The rate for the reaction endothermic by 3 kcal (12·6 kJ) mol^{-1} should be slower by a factor of nearly 150 than the thermoneutral process on the basis of activation energies. It appears that energy released on modification of the molecular geometry in going from ground to triplet states of the acceptor can contribute towards the total excitation energy: the time scale for bimolecular reaction is quite sufficient to allow accompanying skeletal motion. Acceptor molecules having large separations between the (0,0) bands in emission and absorption are therefore most likely to exhibit this behaviour, a view confirmed by such experimental evidence as there is. Where geometrical rearrangement cannot provide the necessary energy, the rate of transfer drops off rapidly as the process becomes more endothermic. Thus, in the transfer of triplet energy to biacetyl, rather than to *cis*-stilbene, the rate of reaction decreases by a factor of about 40 in going from donors for which $\Delta E = 0$ to those for which $\Delta E = +3\,\text{kcal}$ $(12\cdot6\,\text{kJ})$ mol^{-1}. Again, even with cis-stilbene as acceptor, the rate begins to drop off rapidly for ΔE greater than about 5 kcal/mol, which suggests that this is the maximum amount of energy available from geometrical rearrangement.

Energy transfer between complex species has been studied in the gas phase as well as in condensed phases. The system biacetyl–benzene (vapour phase at room temperature) provides an example showing both singlet–singlet and triplet–triplet transfer. Pure biacetyl absorbs strongly at $\lambda = 253\cdot7\,\text{nm}$, but exhibits only a very low yield for emission processes ($< 10^{-4}$ at pressures around 5 torr). At $\lambda = 435\cdot8\,\text{nm}$, both phosphorescence ($\phi_p = 0\cdot15$) and fluorescence are observed. It is believed that S_1 is populated at $\lambda = 435\cdot8\,\text{nm}$, and S_2 at $\lambda = 253\cdot7\,\text{nm}$: photodissociation is the predominant fate of S_2, while fluorescence, or more especially phosphorescence via ISC to T_1, are possible fates of S_1. Addition of benzene to biacetyl irradiated at $\lambda = 253\cdot7\,\text{nm}$ leads to the appearance of visible emission whose spectrum shows it to be biacetyl phosphorescence (and not fluorescence). The enhancement of emission is due to energy transfer from triplet benzene, and not to collisional stabilisation by the benzene of 'hot' biactyl (which might otherwise dissociate). This latter point is established by the ineffectiveness of cyclohexane in increasing the emission yields, even though cyclohexane and benzene have almost equal stabilising efficiencies (at wavelengths at which benzene does not absorb). In fact, at $\lambda = 253\cdot7\,\text{nm}$, the presence of benzene actually increases the quantum yields for the decomposition of biacetyl, and it appears that singlet–singlet energy transfer takes place in addition to the sensitisation of phosphorescence. The second excited singlet, S_2, is probably populated in the singlet energy exchange since no sensitised fluorescence is observed: fluorescence is observed when S_1 is populated by non-sensitised biacetyl absorption at $\lambda = 435\cdot8\,\text{nm}$. Figure 5.3 shows the energy-level diagram used by Ishikawa and Noyes in the interpretation of their experimental results (J. Chem. Phys. **37**, 583 (1962)). Analysis of a reaction scheme based on the paths in the figure predicts exactly the observed kinetic dependence on biacetyl concentration of sensitised phosphorescence yield. The scheme assumes that IC($S_1 \rightsquigarrow S_0$) and ISC($T_1 \rightsquigarrow S_0$) are slow in benzene. If the assumption is correct, we can further calculate the sensitised phosphorescence yield, ϕ_p^s, from ϕ_f for benzene and ϕ_p for biacetyl. At low pressures the singlet–singlet energy transfer process will be slow relative to $ISC(S_1 \rightsquigarrow T_1)$ in benzene, and we may then write, for benzene,

$$\phi_{ISC} = 1 - \phi_f \qquad (5.33)$$

$\phi_f = 0.22$ ($\lambda = 253.7$ nm, $P = 20$ torr, $T = 29°C$), so $\phi_{ISC} = 0.78$. The only path we have left open to triplet benzene is energy

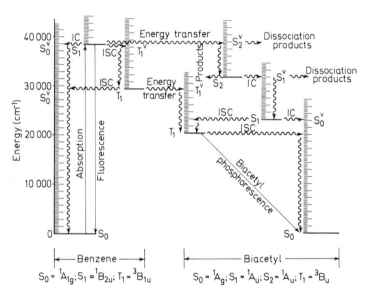

Figure 5.3. Energy level diagram for the system benzene biacetyl showing the several electronic transitions, energy transfer processes and routes to dissociation. (Based on reaction scheme of H. Ishikawa and W. A. Noyes, Jr., J. chem. Phys. **37,** 583 (1962) and reproduced from J. G. Calvert and J. N. Pitts, Jr., Photochemistry, Wiley, New York (1966))

transfer; 0.78 is therefore also the quantum yield of triplet biacetyl formation. ϕ_p(biacetyl) $= 0.15$ in the unsensitised system, so that the sensitised yield, ϕ_p^s, should be $0.15 \times 0.78 = 0.12$. The observed limiting low-pressure yield is 0.120.

5.3 LONG-RANGE, COULOMBIC, INTERACTIONS

There are some energy transfer processes for which the collision diameter is apparently greater than the sum of gas-kinetic collision radii. For example, energy transfer between excited singlet states of hydrocarbons occurs as fast as spontaneous decay at concentrations in benzene around 10^{-2}–10^{-3} mol l^{-1}, which corresponds to a radius, r, between exchanging molecules of about 50Å. Again, the measured rate constants for transfer of excitation in the hydrocarbons seem greatly to exceed the diffusion limited rate, and do not depend on solvent viscosity. Thus rate constants for the process

1-chloroanthracene(S_1) + perylene(S_0) $\rightarrow$ 1-chloroanthracene(S_0)

$$+ \text{perylene}(S_1) \qquad (5.34)$$

are about $1 \cdot 5 \times 10^{11}$ l mol^{-1} s^{-1} in room temperature solutions in benzene or in liquid paraffin, and virtually the same in a glass at $-183°C$.

Although the evidence described in the last paragraph appears to support single-step long-range transfer, it should perhaps be noted that it is not unequivocal. The intermolecular distances calculated from concentrations are *average* values, and may not refer to the distance for the exchanging pair. If the solution deviates from ideality to a sufficient degree, semi-aggregation of solute molecules may invalidate both diffusion rate and intermolecular distance calculations. The situation is aggravated by the lack of confirmatory evidence for long-range transfer in gas-phase experiments. According to the theory for dd-Coulombic interaction between atoms, the probability of energy transfer per gas-kinetic collision should exceed unity if ΔE is less than about 130 cm^{-1}. Unfortunately, no gas-phase energy exchange process has been studied in which $\Delta E < 130$ cm^{-1} and in which the transitions in both partners are fully allowed for electric dipole interaction. The theory has not, therefore, been tested, and some doubt must remain. However, more detailed predictions of the theory of Coulombic interaction in *solution* do accord with experimental results, as we shall see in the rest of this section.

Förster has derived an equation for the rate per second, k_e, of energy transfer by the dipole–dipole *inductive resonance*, or Coulombic interaction, mechanism. The equation reduces to the approximate form

$$k_e \sim 1 \cdot 25 \times 10^{-25} \frac{\phi_D^f}{n^4 \tau_D r^6} \int_0^\infty f_D(\bar{v}) \varepsilon_A(\bar{v}) \frac{d\bar{v}}{\bar{v}^4} \qquad (5.35)$$

where r is in cm. ϕ_D^f is the fluorescence quantum yield for the *donor*, $f_D(\bar{v})$ the normalised spectral distribution (in quantum units) of the emission and τ_D its mean lifetime. $\varepsilon_A(\bar{v})$ is the molar decadic extinction coefficient of the *acceptor* at a frequency $\bar{v}$ in cm^{-1}; n is the solvent refractive index.

We may define a 'critical distance', r_0, as that intermolecular separation at which energy transfer and spontaneous decay are equal: that is, when $k_e = 1/\tau_D$. From Eq. (5.35) we calculate r_0 to be given by

$$r_0 \sim \left[1\cdot 25 \times 10^{-25} \frac{\phi_D^f}{n^4} \int_0^\infty f_D(\bar{v}) \varepsilon_A(\bar{v}) \frac{d\bar{v}}{\bar{v}^4} \right]^{\frac{1}{6}} \qquad (5.36)$$

Table 5.5 compares values of k_e and r_0 calculated from Eqs. (5.35) and (5.36) with some experimental values. The agreement between

Table 5.5. Comparison of experimentally determined exchange rate constants and critical distances with those calculated from Eqs. (5.35) and (5.36). Data from F. Wilkinson, *Adv. Photochem.* **3**, 241, Table III (1964)

Donor	Acceptor	$10^{-10}k_e(l\ mol^{-1}s^{-1})$		$r_0(\text{Å})$	
		Theory	*Experiment*	*Theory*	*Experiment*
Anthracene	perylene	2·3	12	31	54
Perylene	rubrene	2·8	13	38	65
9,10-Dichloroanthracene	perylene	1·7	8·0	40	67
Anthracene	rubrene	0·77	3·7	23	39
9,10-Dichloroanthracene	rubrene	0·85	3·1	32	49

observed and predicted values of k_e is fair; theory and experiment concur well in deriving values for the critical distances, which are considerably in excess of the kinetic collision radii.

Although electric quadrupole interactions with light are weaker by a factor of 10^7-10^8 than electric dipole interactions, the probability of dq energy-transfer (i.e. transfer in which the transition for one partner is dipole-allowed and for the other quadrupole-allowed) is only less by a factor of about 10 than for dd interaction. The theory for the dq process predicts that the strength of interaction should fall off as $1/r^8$, so it is less likely than dd exchange to lead to long-range transfer.

According to the spin selection rule, $\Delta S = 0$, long-range Coulombic transfer should be impossible for any process involving multiplicity changes, and long-range triplet–triplet energy transfer would then be excluded. However, to the extent that spin–orbit coupling allows electric dipole optical transitions with $\Delta S \neq 0$ in complex molecules, Coulombic transfer can occur by the dd mechanism. Transfer is likely to be slower than for exchange processes in which transitions for donor and acceptor are fully allowed, but, since the actual radiative lifetimes of the triplet states are also long, the long-range energy transfer process may still be important relative to radiation. It follows that the long-range interaction is likely to be demonstrated only in systems in which quenching or ISC is not the predominant loss process for the donor triplet. A very interesting possibility arises that a process such as

$$D^*(T_1) + A(S_0) \longrightarrow D(S_0) + A^*(S_1) \qquad (5.37)$$

can take place *more* readily than triplet–triplet transfer. The optical transition is only weak for *one* of the exchanging pair, instead of for both, and if radiationless decay of $D^*(T_1)$ is slow, reaction (5.37) can occur via Coulombic interaction. In contrast, reaction (5.37) is not favoured by the exchange interaction mechanism since it is non-adiabatic, while triplet–triplet exchange is adiabatic. Triplet–singlet exchange has, in fact, been detected by the emission of *sensitised delayed fluorescence* in rigid glasses of triphenylamine (donor) and chrysoidine (acceptor). The experimental value for r_0 is 55Å, and the theoretical value 40Å. Similar fluorescence sensitised by the triplet of triphenylamine has been detected with chlorophyll and pheophytin as acceptors, and such energy transfer may prove to be implicated in photobiological processes.

5.4 MIGRATION OF EXCITONS

The last two sections have discussed the transfer of energy between effectively isolated molecules, in which the electronic energy difference, ΔE, between the states D* and A* is generally much greater than the interaction between initial and final states D*A and AD*. Further, the electronic excitation energy of the individual species is much greater than the molecular interaction energies.

In molecular aggregates or crystals, coupling between molecules may occur as a result of increased interaction energies. The electronic excitation may then be thought of as an exciton which is not bound to any one molecule, and energy transfer from a 'host' to a 'guest' molecule can occur by migration of the host exciton until it meets a guest molecule which acts as an energy 'trap'. Long-range transfer can therefore take place even if the interactions are what would amount, in the isolated molecule, to exchange interactions.

The exciton theory successfully explains many features of absorption and emission spectra in crystals. For so-called 'weak' coupling, the spectra are similar to those of the isolated molecule, although there may be some splitting of the vibrational structure; 'strong' coupling is defined as the case where the spectral transition band in the aggregate is wider than that in the isolated molecule. The extent of splitting is related to the rate of transfer of excitation between individual molecules (i.e. to the rate at which the excitation moves over each molecule in the lattice). These rates may be 10^{12}–10^{13} s^{-1} in a weakly coupled system, and as high as 10^{15} s^{-1} in strong-coupling cases. A splitting of 1 cm^{-1} may typically correspond to about 1000 molecule-to-molecule excitation exchanges in the lifetime of the exciton. Triplet exciton interaction in perdeuterobenzene (doped with C_6H_6) gives a splitting of 12 ± 1 cm^{-1} which corresponds to 10^{12} nearest-neighbour transfers of triplet energy. It is clear that in crystals of this kind the phosphorescent emission can be affected by exceedingly small concentrations of impurity.

Experimental evidence also suggests that large numbers of lattice sites can be visited by an exciton during its lifetime. It is well known that the green fluorescence of anthracene is actually sensitised fluorescence of naphthacene present as a trace impurity. Measurements of the naphthacene concentration suggest that about 10^6 lattice sites are explored during the lifetime of the excited state. The exciton-migration mechanism for energy transfer is confirmed by the absence of sensitised fluorescence from fluid solutions of ordinary anthracene.

Random walk treatment of diffusion gives the distance d travelled by an exciton in a time τ as

$$d = \sqrt{2D\tau} \tag{5.38}$$

where D is the diffusion coefficient for the exciton. Estimates of D (and measurements of τ from the emission lifetime) can therefore be tested against experimental values for d. An elegant method for the determination of d in anthracene has recently been described. Layers of *pure* anthracene of different depths were deposited on glass plates, and an 'exciton detector' layer of naphthacene: anthracene, 1:300, was deposited on top of the pure anthracene.

The pure anthracene layer was irradiated, and sensitised naphthacene fluorescence in the detector layer was observed. By varying the thickness of the anthracene layer it was possible to measure the diffusion length of the exciton. A value of 460Å was obtained in these experiments, which certainly confirms the long-range nature of the energy transfer. The *triplet* exciton in anthracene, with its longer lifetime, probably has a diffusion length of about 5×10^{-4} cm.

Further evidence for migration distances greater for triplet than for singlet excitons has been obtained in studies using C_6D_6 crystals doped with C_6H_5D and C_6H_6. The intensity of sensitised *fluorescence* from the two guests depends only on their relative concentrations. However, the sensitised *phosphorescence* from C_6H_6 is ten times greater than that from C_6H_5D when $[C_6H_6] = [C_6H_5D]$. Triplet states of both guests should be populated from triplet C_6D_6 at nearly the same rates ($\Delta E = -170\,\text{cm}^{-1}$, -200 cm^{-1}, for C_6D_6–C_6H_5D and C_6D_6–C_6H_6 triplet energy transfer, respectively), and it appears that triplet C_6H_5D can populate triplet C_6H_6 ($\Delta E = -30\,\text{cm}^{-1}$) by triplet exciton migration over the large distances between the different guests.

We should note finally that exciton, rather than material, diffusion may account for high rates of energy transfer in some liquids where solvent–solvent interaction leads to migration of energy first between solvent molecules and then to the solute trap.

5.5   ENERGY-POOLING, AND UPHILL PROCESSES

Energy may sometimes be transferred from an excited species to an acceptor which is already excited, thus raising the acceptor to a higher electronic state; the process may be referred to as energy-pooling.

We have referred already (Section 4.7) to 'dimol' emission from two excited oxygen molecules in the $^1\Delta_g$ state

$$O_2(^1\Delta_g) + O_2(^1\Delta_g) \rightarrow 2O_2 + h\nu_{\lambda = 634\,\text{nm}} \qquad (5.39)$$

The emitting species appears to be the *excimer* $[O_2(^1\Delta_g):O_2(^1\Delta_g)]$, and the process is radiative energy-pooling. Further dimol emissions have been observed in the gas-phase oxygen system from the excimers $[O_2(^1\Delta_g):O_2(^1\Sigma_g^+)]$ and $[O_2(^1\Sigma_g^+):O_2(^1\Sigma_g^+)]$. Since the emission intensity will be proportional to the product of excited state concentrations, and thus to the square of the absorbed intensity, the quantum yield for the 'dimol' emission intensities will be dependent on the absorbed light intensity. A more probable excimer emission is that from a dimer made up of one excited and one unexcited molecule, and is not an energy-pooling process. Such emission is observed in the 'prompt' fluorescence of pyrene at high pyrene concentrations: the violet fluorescence of dilute solutions is replaced by a structureless blue emission which is thought to derive from an excimer:

$$\text{pyrene}(S_0) + \text{pyrene}(S_1) \longrightarrow \text{pyrene}:\text{pyrene}(S_0 S_1) \qquad (5.40)$$

$$\text{pyrene}:\text{pyrene}(S_0 S_1) \longrightarrow \text{pyrene}:\text{pyrene}(S_0 S_0) + h\nu \qquad (5.41)$$

Molecular, rather than radiative, energy pooling has been established in many systems. Triplet–triplet pooling to give an excited singlet state is most common, partly because the relatively long lifetime of excited triplets favours the rare triplet–triplet bimolecular process. The reaction

$$D^*(\text{triplet}) + A^*(\text{triplet}) \longrightarrow D(\text{singlet}) + A^{**}(\text{singlet}) \quad (5.42)$$

is adiabatic with respect to spin, and for many organic molecules the first excited singlet is energetically accessible from two triplets. An interesting example of a non-adiabatic energy-pooling process is the formation of the second singlet of $O_2(O_2(^1\Sigma_g^+))$ in bimolecular collision of $O_2(^1\Delta_g)$ molecules:

$$O_2^*(^1\Delta_g) + O_2^*(^1\Delta_g) \longrightarrow O_2^{**}(^1\Sigma_g^+) + O_2(^3\Sigma_g^-) \qquad (5.43)$$

Emission at $\lambda = 762$ nm (of the forbidden $O_2(^1\Sigma_g^+) \to O_2(^3\Sigma_g^-)$ system) is observed even though reaction (5.43) is spin-forbidden; the forbiddenness is reflected in the low rate constant for the process ($\sim 10^3 \, \text{l mol}^{-1} \, \text{s}^{-1}$).

Both donor and acceptor are usually molecules of the same chemical entity, so that reaction (5.42) provides a means of reaching the singlet state when only triplets are present in the system. Energy-pooling between two triplets is known as 'triplet–triplet quenching' or 'triplet–triplet annihilation' and is another mechanism for the emission of *delayed fluorescence* (see also Section 4.6). For example, in anthracene the decay of fluorescence has two components, one with the normal fluorescence lifetime and the other slow, although the spectral distribution of both components is identical. Suggestions that the delayed component might result from formation and subsequent dissociation of an $AA^*(S_0 S_1)$ excimer (A = anthracene) appear to be ruled out because the total emitted energy during the slow decay depends upon the square of the intensity of light absorbed during the irradiation. The actual excitation mechanism (omitting radiationless decay or quenching steps) seems to be

$$A(S_0) + h\nu \rightarrow A^*(S_1) \tag{5.44}$$

$$A^*(S_1) \rightarrow A(S_0) + h\nu \qquad \text{normal fluorescence} \tag{5.45}$$

$$A^*(S_1) \rightsquigarrow A^*(T_1) \qquad \text{ISC} \tag{5.46}$$

$$A^*(T_1) \rightarrow A(S_0) + h\nu \qquad \text{normal phosphorescence} \tag{5.47}$$

$$A^*(T_1) + A^*(T_1) \rightarrow A^*(S_1) + A(S_0) \quad \text{energy-pooling} \tag{5.48}$$

$$A^*(S_1) \rightarrow A(S_0) + h\nu \qquad \text{delayed fluorescence} \tag{5.49}$$

Reactions (5.45) and (5.49) are, of course, identical, and are written twice to show the sequence of events leading to prompt and delayed fluorescence.

There is a possible alternative mechanism which could account for the observed dependence on intensity of delayed fluorescence yield. The sequence, $A(S_0) + h\nu \rightarrow A^*(S_1) \rightsquigarrow A^*(T_1)$; $A^*(T_1) + h\nu \rightarrow X \rightarrow A^*(S_1) \rightarrow A(S_0) + h\nu$, could give a delayed emission if the intermediate X were long-lived, and [X] would be proportional to I_{abs}^2. This mechanism predicts highest [X] and, hence, highest delayed emission intensity when the stationary concentration of $A^*(T_1)$ is highest. However, in rigid media, where decay of $A^*(T_1)$ is slow and $[A^*(T_1)]$ correspondingly high, the delayed emission is extremely weak,† and it must be concluded that delayed fluorescence in solution is not produced by double excitation.

The kind of delayed fluorescence just described does not show the same dependence on temperature as the thermally activated E-type delayed fluorescence (Section 4.6), and it may be distinguished from it by this means. A better distinguishing feature is the emission intensity dependence on the intensity absorbed, which is first-order in E-type delayed fluorescence. Further, E-type delayed fluorescence has the same decay lifetime as that of the triplet–singlet phosphorescence in the same solution; delayed fluorescence excited by the triplet annihilation mechanism should have a lifetime about one-half of that of the phosphorescence, because of the second-order dependence on triplet concentration.

---

†Triplet–triplet annihilation can occur in *crystals* of anthracene, and gives rise to delayed fluorescence. The energy-pooling probably proceeds via triplet exciton migration, and diffusion length estimates of about 5×10^{-4} cm agree with those from other experiments.

Triplet annihilation delayed fluorescence is sometimes known as *P-type delayed fluorescence* because it is observed in solutions of pyrene. However, delayed fluorescence in pyrene shows an additional feature in that the delayed emission appears to derive mainly from the excimer $PP*(S_0S_1)$ [P = pyrene], while the normal, prompt, fluorescence at moderate concentrations shows both monomer and excimer bands. The explanation appears to lie in the mechanism for the triplet–triplet energy-pooling step. If the singlet excimer is a reaction intermediate (related to the transition state), then radiation may occur before the equilibrium concentrations of excimer and monomer can be set up:

$$P*(T_1) + P*(T_1) \rightarrow PP*(S_0S_1) \rightarrow P_2(S_0S_0) + h\nu \qquad (5.50)$$

and the *delayed* emission will show no component from monomeric excited pyrene, $P*(S_1)$.

Sensitised delayed fluorescence is another process that may be consequent upon energy-pooling. For example, a solution of 10^{-3} mol l^{-1} of phenanthrene containing 10^{-7} mol l^{-1} of anthracene shows quite intense delayed emission from the anthracene. This concentration of anthracene is far too low to show direct P-type delayed fluorescence, and the excitation scheme appears to be

$$P(S_0) + h\nu \rightarrow P*(S_1) \qquad\qquad\qquad\qquad\qquad (5.51)$$
$$P*(S_1) \rightsquigarrow P*(T_1) \qquad\qquad\text{ISC} \qquad\qquad (5.52)$$
$$P*(T_1) + A(S_0) \rightarrow P(S_0) + A*(T_1) \quad\text{energy transfer} \quad (5.53)$$
$$A*(T_1) + A*(T_1) \rightarrow A*(S_1) + A(S_0) \quad\text{energy-pooling} \quad (5.54)$$
$$A*(T_1) + P*(T_1) \rightarrow A*(S_1) + P(S_0) \quad\text{energy-pooling} \quad (5.55)$$
$$A*(S_1) \rightarrow A + h\nu \qquad\qquad\text{delayed fluorescence} \quad (5.56)$$

In this system it is not possible to distinguish energy-pooling between like (reaction 5.54) and unlike (reaction 5.55) species. However, if a compound whose triplet energy is less than one-half of its excited singlet energy is chosen as acceptor, then triplet annihilation $A*(T_1) + A*(T_1)$ cannot populate $A*(S_1)$. If the donor triplet level is sufficiently high $D*(T_1) + A*(T_1)$ *can* produce $A*(S_1)$, and it should be possible to observe sensitised delayed fluorescence produced *only* by mixed triplet annihilation. An example of a suitable donor–acceptor pair is anthracene–naphthacene. Dilute solutions show quite intense delayed emission from naphthacene, although solutions of naphthacene alone show no delayed emission even at higher concentrations.

The energy-pooling process leads to more excitation energy in one product than was present in either reactant, and it is occasionally possible for the radiation ultimately emitted to be of *shorter*

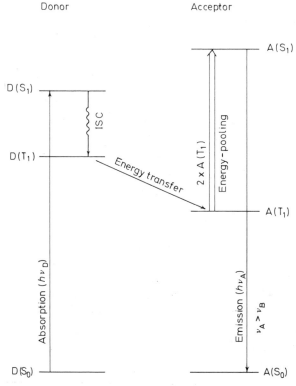

Figure 5.4. The mechanism of excitation of delayed fluorescence in which the emission is of shorter wavelength than the exciting radiation. Note that $D(S_1)$ lies below $A(S_1)$, but $D(T_1)$ lies above $A(T_1)$

wavelength than the exciting radiation. Sensitised 'anti-Stokes' delayed fluorescence has been observed with donor–acceptor pairs such as phenanthrene–naphthalene and proflavine–anthracene. The requirement is that the lowest excited singlet of the donor should lie below that of the acceptor, but that the triplet should lie above that of the acceptor (see Fig. 5.4). The observed anti-Stokes emission from naphthalene has 5 kcal (21 kJ) mol^{-1}, and that from anthracene 9 kcal (38 kJ) mol^{-1}, more energy per quantum than the exciting radiation.

The concentration of the energy from two separately absorbed quanta into a single molecular species is demonstrated most dramatically in the sensitised anti-Stokes emission experiments, although it occurs, in principle, in all energy-pooling processes. The explanation of this apparent contravention of the Planck energy–frequency relation (and of the Stark–Einstein law) may provide an understanding of the primary processes in photosynthesis, in which just such a concentration of photon energy is needed for the occurrence of photochemical reactions.

Photon energy may occasionally be upgraded by true *biphotonic absorption* (as distinct from tandem excitation of one level and then another). The high intrinsic intensity of laser radiation makes possible the 'simultaneous' absorption of two photons, and effects involving double quantum excitation have been observed. For example, emission in caesium vapour from the $9^2D_{\frac{3}{2}} \rightarrow 6^2P_{\frac{3}{2}}$ transition ($\lambda = 584.7$ nm) may be excited by radiation from a laser tuned to $\lambda = 693.78$ nm, although the caesium is transparent to red light of that wavelength under normal conditions. However, $\lambda = 693.78$ nm corresponds to exactly half the energy required to excite the $9^2D_{\frac{3}{2}}$ state of caesium from the ground, $6^2S_{\frac{1}{2}}$, state (28 828 cm^{-1}), and the upper state appears to be populated by two-photon absorption. Similar observations have been made of fluorescence of aromatic hydrocarbons (phenanthrene, anthracene, pyrene, benzpyrene) excited by laser radiation of a wavelength supposedly not absorbed by the molecules. It can be shown that the effect does not result from frequency doubling of the incident radiation, nor from successive excitation. (The latter process is excluded because the lack of absorption at low intensities precludes the formation of the first, intermediate, level; in other systems, tandem excitation can be important, especially if the first level is long-lived). Very high intensity is the requirement for biphotonic absorption, and the properties peculiar to laser radiation (i.e. coherence, monochromaticity and low divergence – see next section) seem not to be essential. Indeed, biphotonic absorption of light from a high intensity xenon flash has recently been demonstrated, and some suggestion has even been found for three-photon absorption.

Although multiphoton absorption processes offer an interesting mechanism for the occurrence of reactions which are at first sight photochemically impossible, it is unlikely that they are significant in any *terrestrial* natural phenomena.

5.6   STIMULATED EMISSION

We now consider radiative loss of excitation in *stimulated* emission processes. *Net* stimulated emission is observed only in systems where the population of the excited state is greater than that of the ground state, a situation described as population inversion (cf. Section 2.3, p. 21). A population inversion with respect to the absorbing state cannot be created by the direct absorption of light, since when ground and upper states reach equal concentrations, the depopulating stimulated emission will be as fast as the populating absorption. Intramolecular or intermolecular energy transfer is an important step in creating the conditions for net stimulated emission, and for this reason the stimulated emission phenomena are treated in the present chapter.

The great stimulus to investigations of population inversion has been the successful operation of lasers, and the principles of laser action must be discussed briefly before we describe methods of achieving the inversion. Figure 5.5 shows the essential features of a

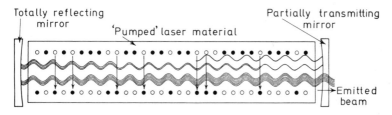

Figure 5.5 Representation of a laser system. Open circles represent the upper state of the emitting species and solid ones the lower state. Induced radiation builds up in intensity as the wave travels to and fro in the cavity between the mirrors. (Based in part on G. C. Pimentel, Scient. Am. (April, 1966))

laser system: the material to be excited in the form of a cylinder (or in a cylindrical tube) is enclosed in an optical cavity. The cavity is designed so that there is constructive interference between forward and reflected waves in each direction. Spontaneous emission initially produces radiation, which passes through the medium and stimulates further excited species to emit. The wave builds up in intensity as it travels back and forth in the cavity, because the emission is radiated in phase with that stimulating it. An output beam is obtained by allowing one endplate to transmit a small fraction of the incident light. Laser radiation possesses several remarkable properties. Because the radiation is emitted always in phase with the other waves in the cavity, the radiation is *coherent*, with respect to both space and time. The coherence is also related to the highly monochromatic nature of the radiation, and to the lack of divergence of the beam. The monochromaticity arises because emission takes place most strongly at the centre of the emission band, and this favoured frequency stimulates further emission at the same frequency, and so on; thus the wave ultimately built up in the laser has a very narrow band width. Off-axis radiation will leave the system after only a few reflections, and will not have an opportunity to build up in intensity; in any case, the cavity will almost certainly not be resonant for the off-axis beam. The emerging beam is therefore virtually non-divergent; it is possible to focus the beam to a spot of very small area, and exceedingly high radiation densities (billions of watts per square centimetre) may be achieved in short-duration pulses. The coherence in time and space of the monochromatic beam leads to interesting interference properties of laser radiation. One of the most striking effects is used in photography by wave-front reconstruction (*holography*). By a suitable technique an interferogram (hologram) produced from an object by the coherent radiation may be recorded on a photographic film without the use of a lens or other image-forming device. Illumination of the hologram with laser radiation leads to reconstruction of the wave fronts, and a three-dimensional image of the original object is seen.

Both pulse and continuous laser action are common. In the first emission ceases when the radiative loss has been sufficient to destroy the population inversion, while in the second the inversion is maintained by the source of excitation. Although the total energy emitted by a laser cannot exceed that provided by the source, and is usually much less, very large *powers* (energy per unit time) may be generated in short-duration pulses. In a typical commercial ruby laser excited by a light flash, the lamp output is 2500 J, and the laser output 25 J delivered in 5×10^{-4} s; the laser power is therefore 50 kW. The technique of Q-switching can enable lasers to deliver even more power by reducing the pulse duration. The laser pulse width is normally a function of the exciting pulse width; if the Q of the optical cavity (the amplification factor) is reduced, possibly by preventing reflection, until the excitation flash is nearly complete, and then restored to its normal value, a laser pulse of very short duration may be obtained. A pulse of 10^{-8} s may be achieved fairly readily, and corresponds in the 25 J laser to a power of 2500 MW. Q-switching may also be used to allow several exciting flashes to be given before laser action is permitted; exceptionally high powers can be produced in this way. Even higher powers may be achieved by a technique known as 'mode-locking'.

Laser action has been described in some detail because ability to make a system 'lase' is probably the best test for the emission of stimulated radiation and, hence, of population inversion. We now turn to photochemical methods of achieving the inversion.

Perhaps the simplest system is the helium–caesium radiative transfer laser. In the ordinary way, gas-phase atomic absorption line widths are too small to allow much energy to be derived from a broad-band source; if, however, a narrow-band source could be found whose emission overlapped the atomic absorption then appreciable excitation could occur. Such a narrow-band source is available to excite the $8^2P_{\frac{1}{2}}$ state of Cs. The He$(3^3P \rightarrow 2^3S)$ transition lies at 388·969 nm, while the $8^2P_{\frac{1}{2}} \leftarrow 6^2S_{\frac{1}{2}}$ absorption in Cs lies at 388·967 nm. By irradiating caesium vapour with light from a helium discharge lamp, a population inversion of the $8^2P_{\frac{1}{2}}$ with respect to $8^2S_{\frac{1}{2}}$ and $6^2D_{\frac{3}{2}}$ states may be achieved. Laser emission of the transitions to these two states has been reported.

It is much more usual to obtain population inversion in gas-phase lasers by energy transfer from a donor excited in an electric discharge. The helium–neon laser provides an example of such a system. Figure 5.6 shows the energy levels for some states of He and Ne: Russell–Saunders coupling does not describe Ne well, and the letters X for the ground state, A, B and C for excited states, are given instead of term symbols. The 2^3S_1 state of He is excited in the electric discharge; it is metastable to radiative decay because the transition to the singlet ground state is quite rigorously forbidden in such a light atom. Instead, energy is transferred collisionally, to excite the C states of neon, whose excitation energy is close to that of He 2^3S_1. There are four sub-levels of the C state which correspond to $^1P_0$, $^3P_0$, $^3P_1$, $^3P_2$ in Russell–Saunders coupling; each is populated efficiently by energy transfer, which exposes the inadequacy of the $^1P_0$ term symbol, since according to this description, the transfer is non-adiabatic. The B state (ten sub-levels) is not populated at this stage, and there exists a population inversion of C with respect to B, so that stimulated emission can be radiated.

A series of lines, of which the strongest is at 1152·3 nm, is obtained. Of course, the B state becomes populated by the radiative decay of C, but the radiative transition B → A, and then decay of A to the ground state, is more rapid than C → B, so that the population inversion may be maintained by continued pumping of the C level

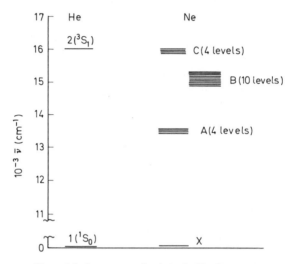

Figure 5.6. Some energy levels in the He–Ne system.

Some such rapid relaxation process is always necessary for continuous laser action.

Another interesting feature of energy transfer is shown in the helium–neon system. The probability for the radiative transition C → X is considerably greater than that for C → B, so that the stimulated emission should be feeble. However, the C → X transition is sufficiently intense for almost all emitted photons to be re-absorbed at the pressures of Ne employed. Successive re-emission and re-absorption steps keep C populated, and it can survive to radiate to the B state.

Solid state lasers are almost always excited by an external source of light;† both flash and continuous versions have been developed. *Intra*molecular energy transfer is important in these lasers. Figure 5.7 indicates the energy levels normally possessed by solid state

†We exclude discussion here of semiconductor lasers, such as the gallium arsenide laser.

laser materials. A suitable host crystal doped with a small proportion (∼0·05–0·1 mol%) of an impurity may show absorption
over a broad band, corresponding to the transition C ← X. Rapid
non-radiative transitions can then populate the narrow level B,
and result in a population inversion of B with respect to A. The
conditions required for laser action are therefore that the band
C be wide enough to absorb an adequate amount of energy from
the exciting light, and that at the same time the transition B → A
be sufficiently sharp for amplification to result. Dopant atoms in
which the optical transitions occur between shielded electrons are
used to obtain the sharp B → A fluorescence; transition metal,

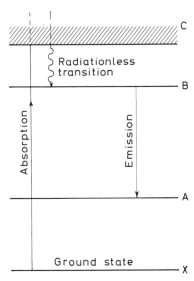

Figure 5.7. Energy levels in a typical
solid state laser material.

lanthanide and actinide impurities have all been successfully employed. Lasers may thus be constructed from materials such as the
fluorides of the alkaline earths doped with samarium or uranium,
or glasses containing neodymium or gadolinium. The most common of all solid state laser materials is, of course, ruby, which is
an aluminium oxide in which a small proportion of Al^{3+} is replaced
by Cr^{3+}. Ruby does not, however, possess distinct levels A and X:
it is a 'three-level system' in which stimulated emission takes place

in transitions to the ground state. This means that more than 50 % of X must be excited (via C) to B before laser action can occur. The rate constants for the several radiative and radiationless processes have been measured: $C \rightarrow X$ ($A_{CX} = 3 \times 10^5 \, s^{-1}$), $C \rightsquigarrow B$ ($k = 2 \times 10^7 \, s^{-1}$), $B \rightarrow X$ ($A_{BX} = 200 \, s^{-1}$). Because of the high rate of the radiationless $C \rightsquigarrow B$ process, depopulation of C takes place rapidly during the exciting flash, and in this way X can lose more than half of its initial population. It is even possible to construct continuously operating ruby lasers, in spite of the unpromising nature of the three-level system; in this case, however, it is necessary to cool the crystal in liquid nitrogen to remove heat liberated to the lattice in the $C \rightsquigarrow B$ step.

Intramolecular energy transfer in metal chelates has been mentioned already (Section 4.5, p. 117); laser action in fluid solution has been observed with materials such as europium benzoylacetonate or dysprosium trisdibenzoylacetonate.

Chemiluminescent systems ought, in principle, to exhibit stimulated emission if a population inversion of excited states could be achieved. *Chemical lasers* currently constitute an active area of research because of the possibility of extremely large energy releases (of the order of $10^3 \, J \, mol^{-1}$) in rapid chemical reactions. Some chemical lasers have, in fact, been constructed. Population inversions between vibrational levels in vibrationally excited HCl can be produced in the $H + Cl_2$ reaction (cf. Eq. 4.39, Section 4.7), and a laser has been made which employs this system. Again, photolysis of methyl iodide can be made to yield a population inversion with respect to the ground state of electronically excited atomic iodine. Stimulated emission of the $I(^2P_{\frac{1}{2}} \rightarrow {}^2P_{\frac{3}{2}})$ transition may then be obtained.

5.7 PHOTOSENSITISED REACTIONS

Intermolecular energy transfer leads to electronic excitation of an acceptor, which can then undergo any of the fates of excited species. One of these fates is decomposition, and *photosensitised reactions* are those in which dissociation or reaction takes place in a species other than that absorbing the radiation.

Sensitisation often makes it possible to induce reaction photochemically in a wavelength region where the reactant is transparent. For example, the first strong absorption in molecular hydrogen is to the $^1\Sigma_u^+$ state, and lies at $\lambda \sim 111$ nm in the vacuum ultra-violet; optical dissociation from $^1\Sigma_u^+$, or $^1\Pi_u$, cannot occur until $\lambda < 85$ nm. However, in the presence of mercury vapour, the mercury resonance line at $\lambda = 253.7$ nm will dissociate H_2 to form initially two ground state hydrogen atoms. The sequence of events can be written

$$Hg + h\nu_{\lambda = 253\ .7\ nm} \rightarrow Hg^* \qquad (5.57)$$

$$Hg^* + H_2 \rightarrow Hg + H + H \qquad (5.58)$$

$Hg^*(^3P_1)$ possesses enough energy (~ 113 kcal (473 kJ) mol^{-1}) to dissociate ground state H_2 to normal atoms (~ 104 kcal (435 kJ) mol^{-1}). As a matter of fact, process (5.58) may not be true energy transfer followed by dissociation, but, rather, chemical reaction,

$$Hg^* + H_2 \rightarrow HgH + H \qquad (5.59)$$

$$HgH \rightarrow Hg + H \qquad (5.60)$$

Although reactions of excited species are discussed in Chapter 6, the mercury-photosensitised process will be treated in the present section. If reaction (5.58) is true energy transfer, it could proceed either to give ground state $H_2(^1\Sigma_g^+)$ with vibrational energy beyond the dissociation limit or to give the unstable state $H_2(^3\Sigma_u^+)$, which also correlates with two ground state H atoms. The first process is formally physical quenching of $Hg(^3P_1)$ and the second energy transfer from it, although there is no difference in the actual quenching effect on the excited atom.

Energy transfer may permit the population of electronic states of molecules different from those populated by absorption, and photosensitised processes may differ chemically from unsensitised photolyses. The direct photolysis ($\lambda < 144$ nm) of methane, for example, yields molecular hydrogen and methylene (CH_2) as the primary products, while the mercury-sensitised photolysis ($\lambda = 253.7$ nm) gives mainly CH_3 and H atoms.

The quantum yield for the direct photolysis at $\lambda = 313$ nm of ethyl pyruvate

$$CH_3COOOC_2H_5 + h\nu \rightarrow 2CH_3CHO + CO \qquad (5.61)$$

is 0.17 in benzene solution. However, in the presence of benzophenone, but at the same wavelength, the quantum yield for ethyl pyruvate removal is 0.32. It is thought that the triplet state of the pyruvate, not rapidly populated from the singlet, is produced by energy transfer from triplet benzophenone, and that it is from this state that dissociation occurs. The transfer is energetically favourable (triplet energies for benzophenone and ethyl pyruvate are 69 kcal (289 kJ) mol^{-1} and 65 kcal (272 kJ) mol^{-1}, respectively); it is interesting that 2-acetonaphthone, with a triplet energy (59 kcal (247 kJ) mol^{-1}) *below* that of the pyruvate, is ineffective as a sensitiser. The energy transfer process is assumed to be adiabatic and, hence, spin-conserved, in systems such as the one described which employ organic sensitisers, and the occurrence of photosensitised reactions provides valuable information about intersystem crossing and triplet chemistry. We shall discuss this further at the end of the section. Spin-conserved energy transfer may also provide a means for the formation of a species of specific multiplicity. For example, the direct photolysis of ketene or of diazomethane yields CH_2 as one of the products; both the triplet ground state and the singlet excited state may be present, the relative proportions depending on the precursor, the pressure and the wavelength of photolysing radiation. However, the benzophenone-sensitised decomposition of diazomethane is thought to give only triplet methylene:

$$B(S_0) + h\nu \longrightarrow B^*(S_1) \overset{\text{ISC}}{\rightsquigarrow} B^*(T_1) \tag{5.62}$$

$$B^*(T_1) + CH_2N_2(S_0) \longrightarrow CH_2N_2^*(T_x) + B(S_0) \tag{5.63}$$

$$CH_2N_2^*(T_x) \longrightarrow CH_2(^3\Sigma_g^-) + N_2(^1\Sigma_g^+) \tag{5.64}$$
$$(B = \text{benzophenone})$$

Similarly, the mercury-photosensitised decomposition of ketene is said, on the grounds of spin-conservation arguments, to yield triplet methylene. These statements should be treated with caution, because the evidence that singlet methylene is not present in a system is often based on the absence of certain reactions assumed to be specific for the singlet state (e.g. retention of stereospecificity in addition to olefins); at the same time, the only *experimental* evidence for the specificity is the non-occurrence of the reactions when it is *thought* that only triplet methylene is present (see Section 6.2 and footnote on p. 172).

The question of spin conservation in collisional energy transfer, and, hence, in photosensitised reaction, is a vexed one, but of importance because of the possibility, touched on in the last paragraph, that specific spin multiplicity could be expected in products of sensitised reactions. Such direct spectroscopic evidence as there is suggests that exchange interaction energy transfer in the gas phase can occur without conservation of spin, mainly because the Russell–Saunders spin, **S**, is not a good quantum number. We saw in the last section, for example, that the energy transfer from He* to Ne that would be spin-forbidden (by Russell–Saunders rules) is just as efficient as the spin-allowed transfers. If **S** is a bad quantum number for neon, it certainly is for mercury, for which the spin-forbidden emission line $^3P_1 \rightarrow {}^1S_0$ is only 100 times weaker than the allowed $^1P_1 \rightarrow {}^1S_0$ line (although it must be said that none of the examples of mercury-sensitised fluorescence given in Section 5.2 are, in fact, spin-forbidden). The same kind of argument applies to sensitisation by organic molecules: radiative and radiationless triplet–singlet transitions occur, yet it must be supposed that **S** is a good enough quantum number to make spin-forbidden collisional *reaction* an inefficient process. However, the assumption of spin conservation is often made, and since experimental results are not often inconsistent with it, we shall imply that spin is conserved in the rest of the discussion. More certainty attaches to experiments where only a triplet state can be populated because of the energies involved (cf. Fig. 5.2 and the associated discussion of sensitised phosphorescence).

Many substances have proved popular as sensitisers. Mercury is a particularly suitable sensitiser for gas-phase experiments because of its volatility at room temperature, and the ease with which emission lamps for the $\lambda = 253 \cdot 7$ nm resonance line can be constructed. Other volatile metals have been used as photosensitisers, among them cadmium, zinc, thallium, indium, calcium, sodium and gallium; the rare gases are useful sensitisers in the vacuum ultra-violet region. Several reports have also appeared of gas-phase reactions sensitised by volatile organic compounds (we have mentioned the decomposition of biacetyl photosensitised by benzene in Section 5.2), and mercury has been used as a sensitiser in condensed phases: in the latter case the resonance line is, of course, broadened to a band. We shall now consider further some aspects of reactions sensitised by metal atoms.

One of the main problems connected with mercury-sensitised reactions is the nature of the transfer step. Physical quenching, accompanied by physical energy transfer, clearly occurs in some systems. Ethylene undergoes photosensitised decomposition to yield acetylene and hydrogen as the main products; secondary reactions involving vinyl radicals are of minor significance, and the mercury-sensitised photolysis of a C_2H_4/C_2D_4 mixture yields only

C_2H_2, C_2D_2, H_2 and D_2, and no HD or mixed acetylenes. It seems, therefore, that the process is

$$Hg^*(^3P_1) + C_2H_4 \longrightarrow Hg(^1S_0) + C_2H_4^* \qquad (5.65)$$

$$C_2H_4^* \longrightarrow C_2H_2 + H_2 \qquad (5.66)$$

It may be that physical energy transfer is possible in this system because it can proceed adiabatically and exothermically to excite the triplet level of ethylene. Mercury-sensitised photolysis of olefins at $\lambda = 184.9$ nm (which populates the singlet, $^1P_1$, state of Hg) appears to yield quite different products: it is suggested that here the upper singlet of the olefin, which can be formed exothermically and adiabatically, is the dissociating state. However, the experimental results are open to question, and, in any case, the effects could be ascribed solely to the greater energy (42 kcal (176 kJ) mol^{-1} more) handed on to the olefin from $Hg^*(^1P_1)$.

Chemical quenching (i.e. via HgH) is believed to be the route to sensitised decomposition of alkanes. Indeed, cadmium can sensitise the decomposition of paraffins, even though the energy of its resonance absorption ($\lambda = 326.1$ nm) is about 10 kcal/mol *less* than that of the R—H bond. It is necessary to invoke the liberation of the 15.5 kcal (65 kJ) mol^{-1} heat of formation of CdH to explain the reaction. Emission of radiation from CdH has, in fact, been observed in some cadmium-sensitised reactions, and there seems little doubt that it is an intermediate in the sensitised decomposition. There is no direct evidence for the formation of HgH, although energy transfer to form the alkane triplet seems to be ruled out on energetic grounds; dissociation from vibrationally 'hot' ground electronic state molecules could still occur.

Even the adiabatic triplet–triplet energy transfer

$$Hg^*(^3P_1) + H_2(^1\Sigma_g^+) \longrightarrow Hg(^1S_0) + H_2(^3\Sigma_u^+)$$
$$\downarrow$$
$$2H(^2S) \qquad (5.67)$$

is energetically possible for the Hg-sensitised photolysis of H_2. One piece of evidence that favours either electronic energy transfer in reaction (5.67) or chemical quenching to HgH in reaction (5.59) is the large cross-section for quenching of $Hg^*(^3P_1)$ by H_2. The value of σ_q^2 measured from fluorescence quenching is 8.6×10^{-16} cm^2, which is nearly as great as the gas-kinetic collision cross-section (about 10^{-15} cm^2); it should be compared with the value of $\sigma_q^2 = 2.7 \times 10^{-17}$ cm^2 obtained for physical quenching by

N_2. Clearly there is some kind of specific quenching in the H_2 case. HgH has recently (1970) been identified spectroscopically as a product of the Hg* and H_2 reaction.

Gunning and his co-workers have used the technique of *monoisotopic sensitisation* to obtain information about the course of mercury-sensitisation in several systems. Natural mercury consists of seven stable isotopes, and the $\lambda = 253 \cdot 7$ nm resonance line consists of ten hyperfine components of varying intensities. By use of a resonance lamp containing only ^{202}Hg, from which a single line is obtained, it is possible to excite that isotope alone in natural mercury vapour in the reaction mixture (the mixture, and more especially the lamp, must be kept cooled so that the width of the ^{202}Hg line is not sufficient to overlap the absorptions of other isotopes). If a product of the sensitised reaction contains mercury, it may have been formed in the primary step or in some secondary step: if the former, the product will show enrichment of the ^{202}Hg isotope; if the latter, the Hg abundances will be normal. Calomel-forming reactions are among those that have been studied by this technique. Mercury-sensitised photolysis of HCl or of alkyl chlorides yields mercurous chloride as a product. Two possible routes can be envisaged:

$$Hg^* + RCl \longrightarrow HgCl + R \qquad (5.68)$$

or

$$Hg^* + RCl \longrightarrow Hg + R + Cl$$
$$Hg + Cl \longrightarrow HgCl \qquad (5.69)$$

The product HgCl contains, in fact, over 50% ^{202}Hg (natural abundance $29 \cdot 8\%$) in some cases. Where RCl is hydrogen chloride the product is enriched by 35%, thus suggesting that 35% of the reaction proceeds via reaction (5.68), and 65% via reaction (5.69).

Monoisotopic sensitisation may offer a method for isotope separation. Mercuric oxide is formed on Hg-sensitised photolysis of water. In the presence of butadiene (to suppress secondary randomising reactions due to OH radicals) up to 95% pure ^{202}HgO can be produced with the ^{202}Hg resonance lamp. Repetitive photolysis could yield ^{202}Hg of high isotopic purity; the other isotopes should be equally amenable to separation.

A very important development in the study of triplet states which depends on photosensitisation has recently been made by Hammond and his co-workers. The technique consists in part of using the rate of an isomerisation of an excited triplet as a measure of the rate at which the triplet is populated (e.g. the rate of *cis–trans* isomerisation in piperylene, 1, 2-dichloroethylene or 2-pentene may be used as an indicator). The triplet of an acceptor may be populated by energy transfer from a donor triplet, and if the singlet of the acceptor lies *above* the donor levels (as in Fig. 5.2), the only excited state of the acceptor that can be populated is its triplet. If $D^*(T_1)$ lies at a sufficiently high energy above $A^*(T_1)$, energy transfer is diffusion-controlled (see Section 5.2), and at moderate A concentrations the rate of the isomerisation is the same as the rate at which $D^*(T_1)$ is populated, so long as intermolecular 'forbidden' singlet–triplet transfer from $D^*(S_1)$ to $A^*(T_1)$ is negligibly slow in comparison. Hence, the rate, or quantum yield, of intersystem crossing in the *donor* $D^*(S_1) \rightsquigarrow D^*(T_1)$ may be calculated.

The method may also be used to obtain an idea of the relative energies in triplet levels of donor and acceptor, and thus fix an unknown triplet energy if the other is known from spectroscopic data. The isomerisation reaction becomes slower as the transfer decreases in efficiency when ΔE for $D^*(T_1)-A^*(T_1)$ becomes positive. The fall in rate may not bear an exponential relationship to positive ΔE (see Section 5.2, p. 142), but it is still possible to obtain a measure of the relative triplet energies.

BIBLIOGRAPHY

A. B. CALLEAR, 'Energy transfer in molecular collisions', in *Photochemistry and reaction kinetics* (Eds. P. G. ASHMORE, F. S. DAINTON, and T. M. SUGDEN), Cambridge University Press (1967)

N. J. TURRO, *Molecular photochemistry*: Chapter 5, 'Electronic energy transfer'. Benjamin, New York (1966)

F. WILKINSON, 'Electronic energy transfer between organic molecules in solution', *Adv. Photochem.* **3**, 241 (1964)

A. L. SCHAWLOW, 'Optical masers', *Scient. Am.* (June, 1961; July, 1963)

O. S. HEAVENS, *Optical masers,* Methuen, London (1964)

G. C. PIMENTEL, 'Chemical lasers', *Scient. Am.* (April, 1966)

H. E. GUNNING and O. P. STRAUSZ, 'Isotope effects and the mechanism of energy transfer in mercury photosensitisation', *Adv. Photochem.* **1**, 209 (1963)

C. A. PARKER, *Photoluminescence of solutions,* Elsevier, Amsterdam (1969)

6

Reactions of excited species

6.1 INTRODUCTION

The chemistry of an excited species may differ markedly from that of the ground state species, and, as we pointed out in Chapter 1, the differences may come about both as a result of the excess energy carried by the excited species and as a result of the particular electronic arrangement of the excited state. Both factors appear clearly in the intramolecular and intermolecular transfer of energy discussed in the last two chapters. Excess energy is obviously a prerequisite for its transfer, and the restriction on the electronic states between which energy may be transferred is a consequence of the manner in which the electrons are arranged in the various states. In the present chapter we shall consider processes, involving excited species, which lead to *chemical* reaction (that is, in which the reactants and products differ in chemical identity rather than state of excitation). These chemical processes can be either intramolecular or intermolecular in the same way as the physical process of energy transfer. The first class of reactions includes intramolecular reductions, additions and various types of isomerisation; intermolecular reactions of excited species include those with added reactants, with unexcited molecules of the absorbing substance, or, in solution, with the solvent.

Much information about the transfer of energy is derived from a study of emission phenomena, and the spectroscopic data implicit in these observations make possible quite precise elucidation of the transfer mechanism. In contrast, the mechanisms of many reactions of excited species are only dimly understood, even though the chemical result is unequivocal. It is necessary, therefore, to consider in this book only those reactions of excited species which exemplify well-substantiated principles. This selection must not be allowed to obscure the very real chemical interest possessed by many processes for which photophysical mechanistic rationalisations have not yet been forthcoming. Photochemical reactions may offer the best synthetic route to a variety of curious and interesting organic compounds: the two examples which follow will serve to illustrate such intermolecular and intramolecular reactions. Irradiation of benzoquinone in the melt leads to cyclodimerisation and the formation of a 'cage' product:

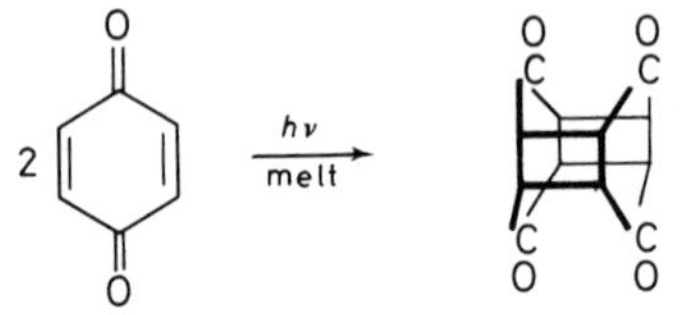

(6.1)

One synthesis of cubane

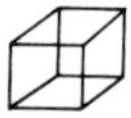

involves an intermediate formed in an *intra*molecular photoaddition of a substituted cyclopentenone moiety to a double bond:

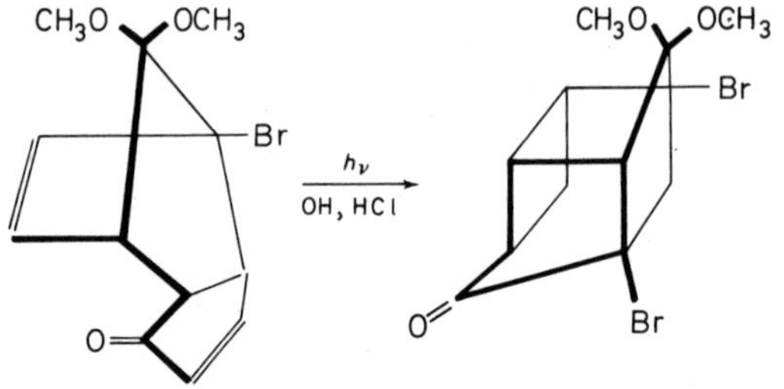

(6.2)

6.2   REACTIVITY OF EXCITED SPECIES

In this section we shall consider in turn three factors that may contribute to the apparent reactivity of an excited species: (1) the intrinsic reactivity of the specific electronic arrangement; (2) the effect of the excitation energy; and (3) the lifetime of the particular excited state.

Atoms or molecules might be expected to show differing reactivity according to the way in which the electrons are distributed in the available orbitals, and, indeed, differences in reactivity for different states can frequently be shown experimentally. For example, following absorption, most aromatic carbonyl compounds undergo rapid intersystem crossing to the lowest triplet. In 'normal' compounds (e.g. benzophenone), this triplet is (n,π^*) in character, although for some *para*-substituted ketones (e.g. *para*-amino-benzophenone) the phosphorescence and ESR spectra suggest that the lowest triplet is either (π,π^*) or a charge-transfer state. The reactions of 'normal' and 'abnormal' compounds are entirely different: triplet benzophenone abstracts H from suitable solvents, and adds across double bonds; triplet p-NH_2-benzophenone does not participate efficiently in either reaction. It is no great surprise, of course, that promotion of a non-bonding electron, essentially localised on the carbonyl O atom, produces a species whose reactivity is not the same as that resulting from promotion of a π electron spread over the $>C=O$ group.

An excited species may be able to participate adiabatically in reactions which are non-adiabatic for the unexcited species. For example, the atom-recombination reactions

$$O + N_2(^1\Sigma) \xrightarrow{M} N_2O(^1\Sigma) \qquad (6.3)$$

and

$$O + CO(^1\Sigma) \xrightarrow{M} CO_2(^1\Sigma) \qquad (6.4)$$

cannot proceed with conservation of spin if the O atom is in its triplet ground state (^3P). Experimental evidence suggests that N_2O or CO_2 are, in fact, produced very inefficiently by $O(^3P)$; excited O in the ^1D state, on the other hand, recombines with N_2 or CO quite readily.† Photolysis of dilute solutions of ozone in liquid or solid N_2 or CO yields atomic oxygen with quantum efficiency $(\phi_{O(^1D)+O(^3P)})$ relatively independent of wavelength in the wavelength range 250 nm to 350 nm. However, the quantum yield for N_2O or for CO_2 formation shows a sharp increase at the wavelength (~ 300 nm) at which $O(^1D)$ becomes an important product of O_3 photolysis.

---

†Although the 'hot' molecule first formed is likely to predissociate in the gas phase; stabilisation is possible in condensed phases.

Excited singlet and triplet states of organic compounds may react in different ways, so long as the chemical reaction proceeds more rapidly than intersystem crossing from S_1 to T_1. For example, the photochemical behaviour of *cis*-dibenzoylethylene is markedly dependent on the multiplicity of the excited state. Direct irradiation in alcoholic solution leads to rearrangement and the formation of an ester of the alcohol:

$$C_6H_5COCH=CHCOC_6H_5 + h\nu \xrightarrow{\ ROH\ } \begin{array}{c} C_6H_5 \\ \diagdown \\ \quad\ C=CHCH_2COOR \\ O\diagup \\ | \\ C_6H_5 \end{array} \tag{6.5}$$

However, the benzophenone-photosensitised reaction, in which the triplet dibenzoylethylene is populated by energy transfer from triplet benzophenone, yields an entirely different product:

$$C_6H_5COCH=CHCOC_6H_5 + h\nu$$

$$\xrightarrow[\phi_2CO\ \text{sens}]{ROH} C_6H_5COCH_2CH_2COC_6H_5 \tag{6.6}$$

Reactions (6.5) and (6.6) appear, therefore, to be specific to the excited singlet and the triplet states, respectively.

The multiplicity of the reacting state can also affect the reactivity in another way. For example, the ground state triplet and first excited state singlet of methylene appear to react in different ways, even though the energy difference between the two states is thought to be small. Absorption spectra in the flash photolysis of diazomethane indicate that $CH_2(^1A_1)$ is first formed, although this bent excited state may be deactivated to the linear ground state $(^3\Sigma_g^-)$ in the presence of excess N_2. Methylene adds to substituted ethylenes to yield substituted cyclopropanes

$$R_1CH=CHR_2 + CH_2 \longrightarrow R_1C\underset{\underset{H_2}{C}}{\overline{\qquad}}CR_2 \tag{6.7}$$

and the addition may be stereospecific. For example, photolysis of mixtures of diazomethane with either *cis*- or *trans*-butene-2 leads to product distributions in the presence of N_2 different from those in its absence: Table 6.1 shows some results.

Table 6.1. Stereospecificity of CH_2 addition to butene-2 (CH_2 from CH_2N_2 photolysis). Data of F. L. Duncan and R. J. Cvetanovic, *J. Am. chem. Soc.* **84,** 3593 (1962); and F. A. L. Anet, R. F. W. Bader and A. Van der Anwera, *J. Am. chem. Soc.* **82,** 3217 (1960)

| | | | | | *% Reaction* | |
					no N_2	*excess* N_2
(a)		$+$	CH_2	Specific	91	17
				Non-specific	9	83
(b)		$+$	CH_2	Specific	91	83
				Non-specific	9	17

The retention of specificity in the absence of N_2 suggests that it is singlet CH_2 that adds stereospecifically; excess N_2 quenches singlet CH_2 to the triplet which adds non-specifically, and the same ratio of

10:2 for ⟨⟩ : ⟨⟩ concentrations is obtained from

both *cis-* and *trans-*olefins. These results are used as evidence that singlet CH_2 adds across the double bond in a single-step process, while triplet CH_2 first forms a diradical, which then ring-closes:

$$CH_2(\text{triplet}) + R_1CH{=}CHR_2 \longrightarrow \left[\begin{array}{c} R_1CH{-}\overset{\bullet}{C}HR_2 \\ | \\ \underset{\bullet}{C}H_2 \end{array} \right] \longrightarrow R_1C{-}CR_2$$

$$(6.8)$$

Rotation about the single C—C bond in the intermediate results in loss of specificity.†

Endothermic reactions require that heat be supplied to the system if the process is to occur spontaneously; if isolated from an external heat source, the system will cool down and reaction will become progressively slower. Even if it is possible to supply energy to the reacting species, the activation energy for reaction must be greater than the heat of reaction. It may be expected, therefore, that highly endothermic reactions will proceed only very slowly at room temperatures. However, the excess energy carried by an excited reactant may be able either to contribute to the kinetic energy needed to overcome the activation barrier, or to participate in a reaction on a different potential surface for which the barrier height between reactants and products is smaller than the ground state activation energy. The reaction of molecular oxygen with ozone is an example of a process which occurs significantly only

Table 6.2. Rate constants (room temperature) for the reaction $O_2 + O_3 \rightarrow 2O_2 + O$.[a] Data from R. P. Wayne, *Adv. Photochem.* **7**, 311 (1969)

State of O_2	Energy above ground state (kcal/mol)	Heat of reaction (kcal/mol)	Rate constant ($l\,mol^{-1}s^{-1}$)
$^3\Sigma_g^-$	0	26 endothermic	$<10^2$
$^1\Delta_g$	21	5 endothermic	$\sim 2 \times 10^6$
$^1\Sigma_g^+$	36	10 exothermic	$\sim 4 \times 10^9$

[a]Original source gives energies in calorie units: 1 cal = 4·184 J.

with excited O_2: Table 6.2 shows rate constants measured at room temperature for the reaction

$$O_2 + O_3 \longrightarrow 2O_2 + O \qquad (6.9)$$

with different states of O_2 as reactant. There is no reason to

†This argument is open to considerable doubt, and is provided solely to illustrate one *possible* influence on reactivity of differing reactant multiplicities. It has been suggested (DeMore and Benson, *Adv. Photochem.* **2**, 258 (1964)) that there is 'as yet no convincing experimental evidence to indicate that addition of triplet CH_2 to olefins is less stereospecific than addition of singlet CH_2'. *Both* singlet *and* triplet CH_2 may first form the diradical, which can then rotate, ring-close or structurally isomerise (propylene is one of the products of the reaction between CH_2 and C_2H_4). Triplet CH_2 is supposedly formed to the exclusion of singlet CH_2 in the benzophenone-sensitised decomposition of CH_2N_2, and *some* non-stereospecific addition is observed; however, a circular argument appears to be used in the proof that triplet CH_2 is involved (see p. 162, Section 5.7).

suppose that the singlet states of O_2 would be intrinsically more reactive towards O_3 by virtue of their electronic arrangement, and the increased rate constants must be ascribed to the excess energy in the excited O_2 molecules.

Even exothermic reactions can possess appreciable activation energies, and energy-rich species might be expected to exhibit enhanced reactivity in such processes. In the few instances where rates of reaction of both ground and excited states have been measured, the energy-rich species is the more reactive. For example, in the decomposition of ozone by atomic oxygen

$$O + O_3 \longrightarrow 2O_2 + O \qquad (6.10)$$

the reaction may be four orders of magnitude faster with excited $O(^1D)$ (process $139 \, kcal \, (582kJ) \, mol^{-1}$ exothermic) than with ground state $O(^3P)$ (process $93 \, kcal \, (389 \, kJ) \, mol^{-1}$ exothermic; $E_a \sim 5 \cdot 6 \, kcal \, (23 \, kJ) \, mol^{-1}$). Since the rate with $O(^1D)$ as reactant approaches the gas-kinetic collision rate, it would appear that the $46 \, kcal \, (192kJ) \, mol^{-1}$ excess energy in $O(^1D)$ can overcome the $5 \cdot 6 \, kcal \, (23kJ) \, mol^{-1}$ activation barrier.

The contributions of intrinsic reactivity and of excess energy to enhanced rates of reaction of excited species could ideally be separated by determination of pre-exponential factors A and activation energies E_a for the reaction. Although the values for A and E_a are not known explicitly for the reaction of $O(^1D)$ in reaction (6.10), the discussion of the last paragraph indicates that $A \sim 2 \times 10^{10} \, l \, mol^{-1} \, s^{-1}$ and $E_a \sim 0$. The A factor for the reaction of $O(^3P)$ is $3 \times 10^{10} \, l \, mol^{-1} \, s^{-1}$, which indicates that the reaction with $O(^3P)$ does not proceed slower than that with $O(^1D)$ because of some factor inherent in the ^3P electronic character. A pair of reactions, for which comparison has recently become possible, is

$$N + O_2(^3\Sigma_g^-) \longrightarrow NO + O + 32 \, kcal \, (134 \, kJ) \, mol^{-1} \qquad (6.11a)$$

$$N + O_2(^1\Delta_g) \longrightarrow NO + O + 54 \, kcal \, (226 \, kJ) \, mol^{-1} \qquad (6.11b)$$

(quoted exothermicities assume ground state N, NO and O). The activation energies for reactions (6.11a) and (6.11b) are $7 \cdot 1 \, kcal$ $(30 \, kJ) \, mol^{-1}$ and about zero, respectively, so that some of the $22 \, kcal \, (92 \, kJ) \, mol^{-1}$ excitation energy of $O_2(^1\Delta_g)$ can apparently contribute towards overcoming the activation barrier. On the other hand, the A factors are about $8 \times 10^9 \, l \, mol^{-1} \, s^{-1}$ and $2 \times 10^6 \, l \, mol^{-1} \, s^{-1}$ for reactions (6.11a) and (6.11b), respectively, and the intrinsic reactivity of $O_2(^1\Delta_g)$ towards ground state

N atoms is *less* than that of $O_2(^3\Sigma_g^-)$, even though the reaction of the ground state O_2 is actually slower at room temperatures.

Several of the examples given of reactions involving energy-rich species have been perverse in the sense that the less reactive *ground* states have been the triplets of O, CH_2 and O_2, while the state exhibiting enhanced reactivity is the singlet. In most organic molecules, the ground state is a singlet, and the triplets are necessarily energy-rich. There is, in fact, a general belief that the triplet state of organic molecules is the one which always participates in excited state reactions because of its high reactivity. This view seems, however, to be ill-founded. Although there may sometimes be a definite dependence of reaction path on multiplicity, as in some of the examples cited earlier in the section, in many instances both singlet and triplet participate in the same reactions. Further, the excited singlet *may* be more reactive than the triplet. Thus it would, perhaps, be more accurate to say that in many photochemical reactions of organic species the triplet is not innately more reactive than the singlet, but it contributes more than the excited singlet to the overall reaction. Even if a triplet is less reactive than a singlet state, it may survive radiative, radiationless and collisional quenching so much better than the singlet that it has a greater probability of undergoing the chemical reaction. In other words, the rate constant for reaction *relative* to that for loss processes may be higher for T_1 than for S_1, even if the absolute rate constant for reaction is lower for the triplet. Kinetic arguments may, in fact, be used to 'prove' the participation of the triplet state. For example, let us consider the photoreduction of benzophenone in the presence of a suitable hydrogen donor, RH,

$$C_6H_5COC_6H_5 + h\nu \xrightarrow{\ \ RH\ \ } C_6H_5C(OH)C_6H_5 \qquad (6.12)$$

(This photoreduction reaction will be described further in Section 6.5.) With a hydrogen donor such as benzhydrol, $C_6H_5CH(OH)$ C_6H_5, at $0.1\ \text{mol}\,l^{-1}$ concentration, the quantum yield for benzophenone disappearance, ϕ_B, is nearly unity. This at once excludes the excited singlet of benzophenone ‹as the reactive species. The rate constant for H atom abstraction by the singlet must be less than $10^8\ \text{l}\,\text{mol}^{-1}\,\text{s}^{-1}$, since physical quenching of S_1 proceeds at least 100 times faster, and yet is diffusion-controlled (i.e. $k_q \not> 10^{10}$ $\text{l}\,\text{mol}^{-1}\,\text{s}^{-1}$). At the same time, the rate constant for $S_1 \rightsquigarrow T_1$ intersystem crossing in benzophenone is about $10^{10}\ \text{s}^{-1}$, so that competition between hydrogen abstraction and ISC places a limit on ϕ_B of 10^{-3} at $[RH] = 0.1 \text{ mol } 1^{-1}$ for reaction of S_1. On the other hand, loss processes for T_1 are much slower than those for S_1 (e.g. the rate constant for $T_1 \rightsquigarrow S_0$ ISC in benzophenone is about 10^5 s^{-1}), and reaction competes effectively with the other processes. Further confirmation that the triplet is the important reactive species comes from a comparison of rate data obtained explicitly for the triplet with that obtained, from the kinetic dependences of ϕ_B, for the 'unknown' state involved in the photoreduction. The benzophenone triplet has been identified in flash photolysis experiments, and the individual rate constants for quenching, ISC and H atom abstraction are virtually identical with those derived for excited benzophenone from the data for ϕ_B; hence, the excited benzophenone is very probably in the triplet state.

In this section we have mentioned some general considerations governing the reactivity of excited species. We shall now mention a few of the more important specific types of intramolecular and intermolecular processes which have been observed.

6.3 INTRAMOLECULAR PROCESSES: ISOMERISATION AND REARRANGEMENT

A wide variety of isomerisations and rearrangements can be induced photochemically. The use of *cis–trans* isomerisation as an indicator of triplet energy transfer has already been referred to several times in Chapter 5. The occurrence of such isomerisation may be used to establish the energy of triplet levels in donors and to determine the quantum yield of intersystem crossing $S_1 \rightsquigarrow T_1$ in donor molecules (p. 166). The availability of the rearrangement energy in suitable acceptors (e.g. *cis*-stilbene) has also been mentioned (p. 142) in connection with the activation energies of some endothermic energy transfer processes. A more complete discussion of these topics is given in the review by Wagner and Hammond (Bibliography to this chapter); here we shall discuss the *cis–trans* isomerisation process itself, rather than pursue further the applications of the 'chemical' technique for elucidation of photophysical mechanism. We shall also consider briefly some examples of structural and valence isomerisation brought about by absorption of light.

Ethylene and its derivatives have lowest excited singlet and triplet states which are (π, π^*) in character: an electron is promoted from the highest filled bonding π orbital to the lowest antibonding π^*

orbital. Both singlet and triplet (π, π^*) states are possible, the state of high multiplicity being of lower energy. At the same time, it can be shown that the (π, π^*) state is most stable if the molecule is twisted, from the planar ground state configuration, through 90° about the double-bond axis. This perpendicular configuration minimises the overlap between π and π^* orbitals, and the T_1 state has its lowest energy for the 90° twist. Figure 6.1 indicates the energies of lowest singlet and triplet states of ethylene as a function of angle of twist. It is apparent that if an olefin is excited to the

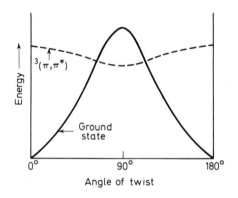

Figure 6.1. Variation with angle of twist between CH$_2$ groups of potential energy of the lowest singlet and triplet states of ethylene. (Based on a diagram of R. S. Mulliken and C. C. S. Roothan, Chem. Rev. **41**, 219 (1947))

(π, π^*) state, then it will tend to twist to the perpendicular configuration. Subsequent electronic energy degradation to the ground state will then require that the molecule become planar again, both *cis*- and *trans*-isomers being formed. The perpendicular form of the triplet is geometrically equivalent whether it is derived from a *cis*- or a *trans*- ground state molecule. It is the increased stability of the perpendicular configuration which sometimes makes the energy difference $S_0 - T_1$ obtained from absorption spectra incompatible with the energy for T_1 suggested by the rates of energy transfer (cf. p. 142). Phosphorescence from olefin triplets is not observed, and this is also probably a result of the relaxation of the triplet to a perpendicular form: the geometries of excited and ground states are so different that Franck–Condon restrictions effectively forbid emission to the lower vibrational levels of the ground state.

Direct evidence that the triplet state can isomerise has been obtained in experiments with the 1, 2-dichlorethylenes. The transition probability of the $T_1 \leftarrow S_0$ absorption can be increased by the presence of the paramagnetic molecule O_2, and absorption of near-ultra-violet radiation in the $T_1 \leftarrow S_0$ band causes isomerisation. The quantum yields for isomerisation suggest that the planar triplets, first formed from *cis*- or *trans*- ground state molecules, convert to a common (perpendicular) triplet much more rapidly than they can be quenched. The common triplet then decays to the *cis*- or *trans*- ground state in the ratio 4:3, regardless of the isomer initially present. Similar evidence for the participation of a common triplet is provided by the equal probability of *cis*- and *trans*-butene-2 formation in the isomerisation of either isomer of butene-2 (triplet-)photosensitised by benzene.

Geometrical photoisomerisation can occur from *cis*- to *trans*- or from *trans*- to *cis*-isomers, so that on prolonged irradiation of either isomer, a photostationary state is set up. If the quantum yields for processes in each direction are similar, the concentrations of *cis*- and *trans*-isomers at the stationary state are determined mainly by the extinction coefficient of each isomer at the wavelength employed. For many simple olefinic systems, the higher extinction coefficient at longer wavelengths is possessed by the *trans*-isomer, and if long wavelength light is used, the ·*cis*-isomer will predominate at the photostationary state.

The photostationary concentrations in a *sensitised* isomerisation might be expected to depend solely on the rate of conversion of the common triplet to the two ground state isomers. With sensitisers whose triplet energy lies well above that of the olefin, this expectation is borne out. For example, the sensitised geometrical isomerisation of 1,3-pentadienes gives a photostationary mixture containing 55% of the *trans*-compound when the sensitiser triplet lies above about 62 kcal (259 kJ) mol^{-1}, and this percentage remains unaltered for sensitisers transferring larger amounts of triplet energy (up to the 78 kcal (326 kJ) mol^{-1} possessed by pyridine). Thus the transfer of energy appears to occur on every collision, with both *cis*- and *trans*-acceptors, and the ratio of the two isomers at the stationary state is determined by the relative rates at which they are formed from the perpendicular triplet. However, for sensitisers with triplet energies *below* 62 kcal (259 kJ) mol^{-1}, increasingly large fractions of the *trans*-isomer appear in the photostationary mixtures (up to 80% for $E_T \sim 57$ kcal (238 kJ) mol^{-1}). Transfer is, therefore, appreciably more efficient to the *cis*- than to the *trans*-isomer at these energies. At even lower sensitiser energies $(E_T < 53$ kcal (222 kJ) $mol^{-1})$ little isomerisation occurs at all, and the 'effective' energy of the perpendicular triplet probably lies at around 53 kcal (222 kJ) mol^{-1} above the *cis*-ground state. The inefficiency of energy transfer to the *trans*-relative to the *cis*-compound for $53 < E_T < 62$ suggests that the excitation energy (to the perpendicular triplet) is greater for *trans*-than for *cis*-isomers, a result, perhaps, not unexpected (the oxygen-induced (0,0) $T_1 \leftarrow S_0$ absorption bands $- T_1$ presumably planar $-$ are at 59 kcal (247 kJ) mol^{-1} and 57 kcal (238 kJ) mol^{-1} for *trans*- and *cis*-isomer, respectively).

Selective stabilisation of the *excited trans*-singlet has been suggested as an explanation of relatively large activation energies found for the *trans*- $\rightarrow$ *cis*-process, but not for the *cis*- $\rightarrow$ *trans*-conversion, in the unsensitised photoisomerisation of stilbene, of some of its 4,4'-derivatives and of azobenzenes. The activation energy for the *trans*- $\rightarrow$ *cis*-isomerisation is markedly dependent on the polarity of the solvent used, and it is postulated that the highly polar S_1 state of the *trans*-isomer is stabilised by solvent interaction. The *cis*-S_1 state, on the other hand, is not stabilised to anything like the same degree, and it can pass over to the T_1 state with virtually no activation energy.

Photochemical valence and structural isomerisations are well known, although the mechanisms are not generally well understood. The ring-closing reactions of dienes and trienes are typical of the valence isomerisations. Photolysis of butadiene (in ethereal solution) yields cyclobutene (and a little bicyclobutane):

$$(6.13)$$

There is much evidence to support the view that the isomerisation reactions of polyenes are different for excited singlet and triplet states. For example, the direct photolysis of 3-methylene-1,5-hexadiene yields mainly 1-allylcyclobutane, while in the presence of a triplet sensitiser only 2-methylenebicyclo [2.1.1]hexane is formed:

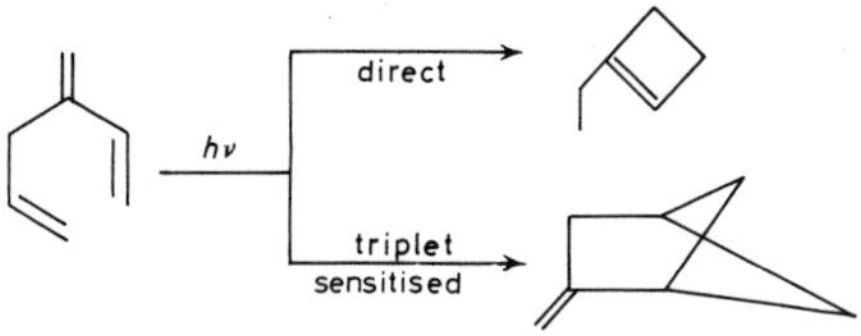

(6.14a)

(6.14b)

Since the products of reaction (6.14b) are probably formed via the triplet of the triene, it may be inferred that the excited singlet is responsible for the entirely different products of the direct photolysis. An exactly similar pair of reactions is observed in the photochemistry of myrcene (formula below); the main products are indicated in the equations:

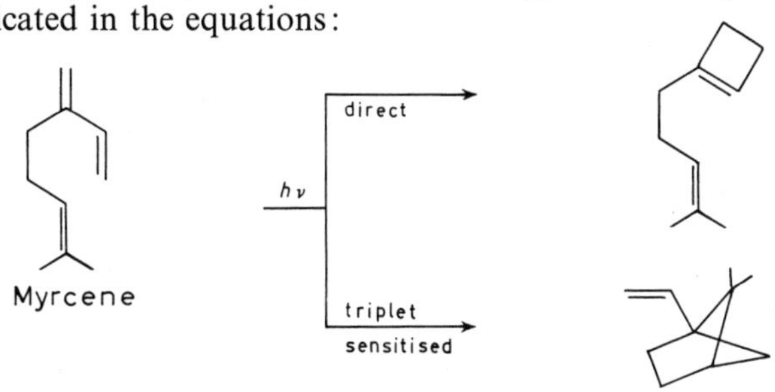

(6.15a)

(6.15b)

In both direct photolysis reactions (6.14a) and (6.15a), a *small* amount of the triplet product is detected, and probably arises from a triplet formed by intersystem crossing from the excited singlet. The strong preference for the type of cyclisation in (6.14b) and (6.15b) supports a hypothesis that cycloaddition reactions of triplets involve two steps with the intermediate formation of biradicals.

Although the general rule is that different products result from sensitised and unsensitised photolyses of dienes, there are some exceptions. For example, 1,1′-bicyclohexenyl and 1,3-cyclooctadiene give cyclobutene derivatives both on direct and on sensitised photolysis:

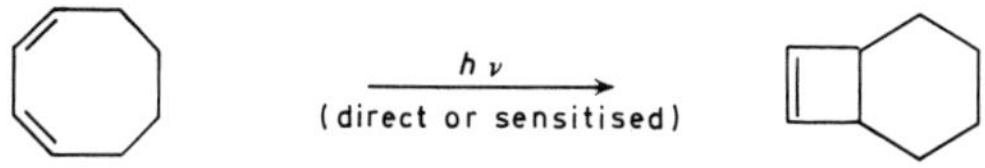

(6.16)

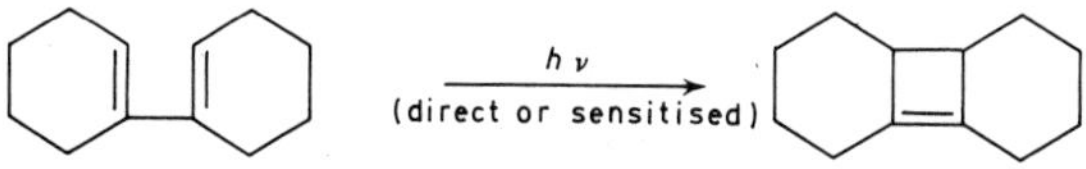

(6.17)

It is possible to predict the stereochemical course of reactions where cyclisation involves bond formation between terminal carbon atoms of a linear conjugated chain. The assumption is made that the new bond involves overlap of portions of the excited π^* orbital associated with the terminal carbon atoms; the *bonding* overlap must involve wave functions of the *same* sign, and this factor determines the final stereochemistry. For example, consider the substituted *cis*-hexatriene

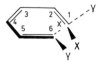

The π^* wave function possesses opposite signs at C_1 and C_6 and overlap of positive or negative portions will require specific rotation, and thus final orientation of the substituents. This statement can be clarified by drawing the individual C_1 and C_6 p orbitals rather than the π^* orbital:

$$\text{(6.18)}$$

Finally, in this discussion of photoisomerisation, we must give an example of a structural isomerisation process, in which the position of atoms changes. Many dienones undergo unsensitised structural photoisomerisation:

$$\text{(6.19)}$$

The reaction can be sensitised by (triplet) acetophenone; it is therefore proposed that the direct photolysis (6.19) involves absorption to S_1 and ISC to T_1 (note, however, that the triplet energy of acetophenone — 74 kcal (310 kJ) mol^{-1} — is enough to excite S_1 of the dienone, which lies at 73 kcal (305 kJ) mol^{-1}). The excited state probably forms a new bond

$$(6.20)$$

and it is thought that the final products of reaction (6.19) may be derived via an ionic intermediate.

6.4 INTRAMOLECULAR PROCESSES: HYDROGEN ABSTRACTION

Hydrogen abstraction is one of the most important intramolecular reactions of excited species. The process is typical of molecules possessing lowest excited states (n,π*) in character (e.g. aldehydes and ketones); indeed, those 'abnormal' carbonyl compounds whose lowest excited levels are (π,π*) states (cf. Section 6.2, p. 169) undergo neither intramolecular nor intermolecular H-abstraction. The intramolecular abstraction of H is of particular importance in the photochemistry of most carbonyl compounds, since it is part of the sequence of events leading to the 'Norrish Type II' fragmentation (Section 3.6, p. 73). Equation (3.59) shows a six-membered transition state in the Type II fission of a ketone: this cyclic intermediate favours intramolecular H-abstraction over intermolecular abstraction from the solvent, and the individual steps of Eq. (3.59) may be written more fully as

$$(6.21)$$

The enol of acetone has been explicitly detected in this reaction. The rate of enol disappearance is the same as the rate at which acetone is formed, a result which lends support to the proposed mechanism for the Type II reaction.

In addition to the fragmentation reaction of the hydroxy biradical formed by intramolecular H-abstraction, there is a ring-closure path leading to cyclobutanol formation:

$$R-\overset{\overset{\displaystyle OH}{|}}{\underset{\bullet}{C}}-CH_2CH_2\overset{\bullet}{C}H_2 \longrightarrow R-\overset{\overset{\displaystyle OH}{|}}{\underset{\underset{\displaystyle CH_2-CH_2}{|}}{C}}-\overset{\overset{}{|}}{CH_2}$$

(6.22)

The electronic state of the biradical involved in reaction (6.22) is not established. In certain cases, the photochemical cyclobutanol formation appears to be stereospecific, e.g.

$$(CH_3)_2CH-\overset{\overset{\displaystyle O}{\|}}{C}\overset{CH_2-CH_2}{\underset{CH_2=CH}{\diagdown}}\overset{H}{\underset{CH_3}{\diagup}}C \xrightarrow{h\nu} (CH_3)_2CH-\overset{QH}{C}\overset{CH_2-CH_2}{\diagdown}\overset{}{\underset{CH=CH_2}{C}}-CH_3$$

(6.23)

The retention of configuration suggests either that ring closure is so rapid that rotation about the C—C bond does not occur, or that a concerted, one-step, four-centre process, involving stereospecific removal of the allyl γ-hydrogen atom, leads to the observed product. If the first mechanism is operative, then a singlet biradical seems to be implicated because of the rapidity of ring formation; on the other hand, cyclobutanol formation *can* proceed via the triplet, since the reaction may be sensitised by (triplet) benzophenone. Reactions (6.21) and (6.22) possibly involve different intermediates, at least in the gas phase, since the relative efficiencies of Type II fission and of cyclobutanol formation depend on experimental parameters such as temperature and pressure. It may be that the relative extent of the two reactions is determined by the degree of vibrational excitation in the biradical, and that the six-membered ring and biradical intermediates are otherwise identical for the fission and ring-closure processes.

6.5 INTERMOLECULAR PROCESSES: HYDROGEN ABSTRACTION

Excited (n, π^*) states, especially those of aromatic carbonyl compounds, not only undergo the intramolecular reduction reactions described in the last section, but can also abstract hydrogen in intermolecular processes. We referred in Section 6.2 to the photoreduction of benzophenone in the presence of a suitable hydrogen donor:

$$C_6H_5COC_6H_5 + RH \xrightarrow{hv} C_6H_5C(OH)C_6H_5 + R \qquad (6.24)$$

The ketyl radicals formed can participate in secondary reactions, which include dimerisation to form a pinacol:

$$
2C_6H_5C(OH)C_6H_5 \longrightarrow C_6H_5 -\!\!\! \underset{\underset{\displaystyle C_6H_5}{|}}{\overset{\overset{\displaystyle OH}{|}}{C}} \!\!\!-\!\!\!-\!\!\! \underset{\underset{\displaystyle C_6H_5}{|}}{\overset{\overset{\displaystyle OH}{|}}{C}} \!\!\!-\!\!\!-\!\!\! C_6H_5 \qquad (6.25)
$$

Photopinacolisation has been recognised since the beginning of the century, when it was found that good yields of benzopinacol were produced by the action of sunlight on solutions of benzophenone.

In a good hydrogen-donating solvent, such as ethanol, the quantum yield for benzophenone disappearance is near unity,

Table 6.3. Dependence of quantum yield for benzophenone disappearance on nature of solvent. Data of A. Beckett and G. Porter, *Trans. Faraday Soc.* **59,** 2039 (1963)

Solvent	Limiting quantum yield for benzophenone removal
Water	0·02
Benzene	0·05
Toluene	0·45
Hexane	1·0
Ethanol	1·0

Benzophenone concentrations in the range $10^{-4}-10^{-1}$ mol l^{-1}

although in certain solvents, for which the activation energy of H-abstraction is high, the quantum yield may be much smaller. Table 6.3 shows some results for different solvents. Under suitable conditions, quantum yields around *two* are observed for removal of benzophenone in isopropyl alcohol solutions. The radical produced from the solvent is itself capable of reducing a molecule of benzophenone to its ketyl radical, and two molecules of benzophenone are removed for each quantum of light absorbed:

$$(C_6H_5)_2CO + (CH_3)_2CHOH \xrightarrow{h\nu} (C_6H_5)_2COH + (CH_3)_2COH \tag{6.26}$$

$$(CH_3)_2COH + (C_6H_5)_2CO \longrightarrow (CH_3)_2C{=}O + (C_6H_5)_2COH \tag{6.27}$$

Isopropyl alcohol is, in fact, used as a standard hydrogen donor in studies of ketone photoreduction.

The efficiency of photoreduction depends not only on the nature of the solvent, but also on the structure of the ketone. Several *ortho-* and *para*-substituted benzophenones exhibit quantum yields markedly smaller than the yield for removal of the parent compound. Some typical quantum yields for photoreduction in isopropyl alcohol solution are given in Table 6.4.

Table 6.4. Quantum yields for reduction of substituted benzophenones (isopropyl alcohol solution). Data from N. J. Turro, *Molecular photochemistry*, Table 6–4, p. 144, Benjamin, New York (1966)

Ketone	Limiting quantum yield for ketone disappearance
Benzophenone	2·0
o-*Tert*butylbenzophenone	0·5
o-Methylbenzophenone	0·05
o-Ethylbenzophenone	0·02
o-Hydroxybenzophenone	~ 0
p-Methylbenzophenone	0·5
p-Phenylbenzophenone	0·2
p-Hydroxybenzophenone	0·02
p-Aminobenzophenone	~ 0

The decrease of photoreduction efficiency appears to result from different causes, according to whether the substituent is in the *ortho-* or *para*-position. *Ortho*-substitution makes possible *intra*-molecular hydrogen abstraction via a six-membered transition state

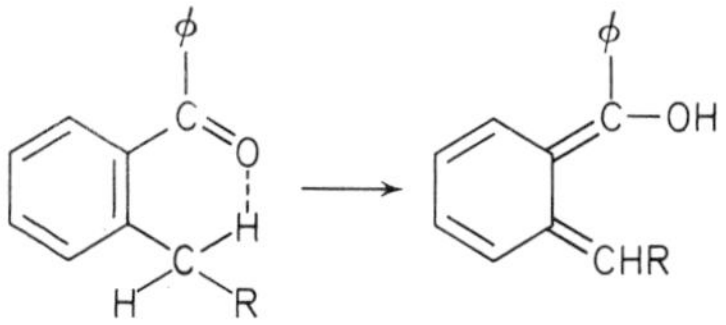

(6.28)

and it is significant that the *o-tert*butyl derivative, which cannot form the six-membered ring, photopinacolises quite efficiently (this result also suggests that steric hindrance is not the main factor in reducing the efficiency of intermolecular reaction of other *ortho*-substituted benzophenones). In a similar manner, both *o*-methoxy- and *o*-carboxy-benzophenone photopinacolise efficiently. Several pieces of evidence corroborate the suggestion that the intramolecular photoenolisation reaction (6.28) quenches the intermolecular reduction. Irradiation of some *ortho*- alkyl substituted benzophenones in CH_3OD leads to introduction of deuterium into the alkyl side chain, presumably via the sequence

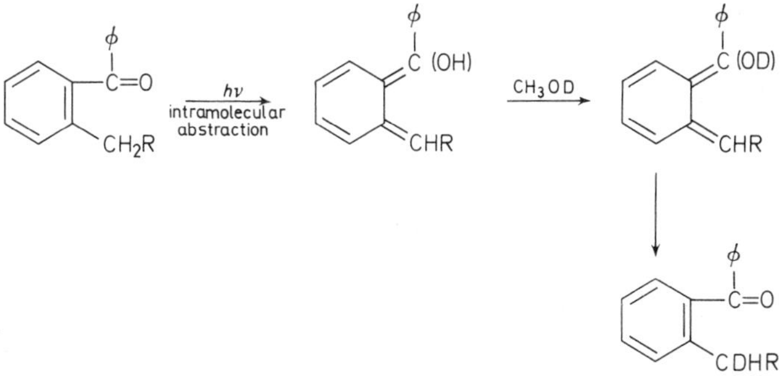

(6.29)

The spectroscopic and kinetic behaviour of a transient absorption, seen in flash photolysis studies, suggests that the transient species is the enol. Further, a Diels–Alder adduct of the enol has been detected on photolysis of *o*-substituted benzophenones in the presence of $CH_3COOC \equiv CCOOCH_3$.

Para-substitution of benzophenone seems to exert its inhibiting effect on intermolecular photoreduction by changing the nature of the excited state. We have referred already to the spectroscopy of these 'abnormal' ketones. The phosphorescence lifetime of p-phenylbenzophenone, for example, is about 50 times longer than that of 'normal' benzophenones, and this, and the structure of the emission spectrum, suggest that the lowest triplet is (π, π^*) in character. A similar conclusion is reached from EPR studies of an irradiated glass containing p-phenylbenzophenone. The (π, π^*) state is likely to be much less reactive with respect to hydrogen abstraction than the (n, π^*) state, and photoreduction of the *para*-substituted ketones will be correspondingly inefficient. An alternative excited state for the benzophenones *para*-substituted by electron-releasing groups is an unreactive intramolecular charge-transfer state, such as

$$\overset{+}{N}H_2 = \underset{}{\bigcirc} - C = \overset{O-}{\underset{}{C}} - \phi$$

which is stabilised by solvation in isopropyl alcohol solution. This latter explanation is able to account for the relatively *high* quantum yields observed for photoreduction in non-polar cyclohexane as solvent (0·9 for p-OH, and 0·2 for p-NH_2 compounds): the lack of solvation leaves the ordinary (n, π^*) state as the lowest triplet in cyclohexane. It is probable that the lowest excited states in isopropyl alcohol have both some (π, π^*) and some charge-transfer character, and the lack of reactivity is a consequence both of delocalisation of $^3(\pi, \pi^*)$ excitation and increased negative charge on the excited carbonyl oxygen.

6.6   INTERMOLECULAR PROCESSES: ADDITION REACTIONS

A wide variety of photoaddition reactions is known, the products of which are frequently cyclic. Both homoaddition and heteroaddition reactions can occur. We shall mention a few of the more important types of reaction.

Conjugated olefins can be photodimerised in the presence of triplet sensitisers. Cyclopentadiene, for example, produces the dimers in approximately equal yields with many ketones as sensitisers, and the relative yields of the three isomers seem not to depend on the nature of the sensitiser. A similar result is obtained for the dimerisation of cyclohexadiene: the isomers

are formed in relative yields of about 60%, 19% and 21% for sensitisers whose triplet energy ranges between 53 kcal (222 kJ) mol^{-1} and 69 kcal (289 kJ) mol^{-1}. Entirely different behaviour is shown by open-chain dienes, such as butadiene. The products of the sensitised dimerisation of butadiene are the cyclobutane derivatives

and, in addition, 4-vinylcyclohexane. At high sensitiser triplet energy (e.g. acetophenone, $E_T \sim 74$ kcal (310 kJ) mol^{-1}), the cyclobutane derivatives are the major products, but with sensitisers whose energy lies below about 60 kcal (251 kJ) mol^{-1}, the fraction of cyclobutanes drops off relative to the vinylcyclohexane. The $S_0 \rightarrow T_1$ energy for the s-trans-rotomer of butadiene is itself about 60 kcal (251 kJ) mol^{-1}, while that for s-cis-butadiene is probably lower (possibly about 54 kcal (226 kJ) mol^{-1}, an energy below which ketones do not efficiently sensitise the dimerisation to any of the products). The different yields of the dimers may, therefore, be a result of the existence of non-interconvertible, stereoisomeric triplet states of open-chain dienes.

Cyclobutane derivatives are also formed by the photoaddition of olefins to the unsaturated bond in α,β-unsaturated ketones. For example, cyclobutane derivatives are the major products of the irradiation of cyclohexenone in the presence of isobutylene

(6.30)

The preferred path of attack seems to involve the carbon α to the carbonyl, and the predominant intermediate is probably

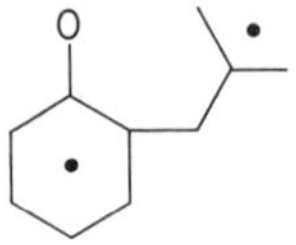

This either ring-closes to yield the cyclobutane derivative, or forms

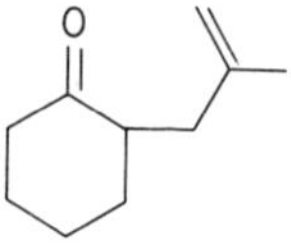

an observed minor (12%) product of the photoaddition. Yields are considerably smaller for the open-chain and cyclobutane compounds derived from the β-substituted intermediate.

Cyclopentenone undergoes similar photoaddition to alkenes, although neither it nor cyclohexenone adds efficiently to dienes. Both ketones can sensitise the dimerisation of dienes, presumably by transfer of triplet excitation, which suggests that the addition proceeds via the triplet of the ketone and that inefficient addition to dienes is a consequence of quenching of the reactive state by energy transfer to the diene.

Carbonyl compounds possessing α,β-unsaturation can, of course, add to each other. Photo(cyclo)dimerisation is known for several such carbonyl compounds in condensed phases; perhaps the best-studied reaction is the dimerisation of coumarin:

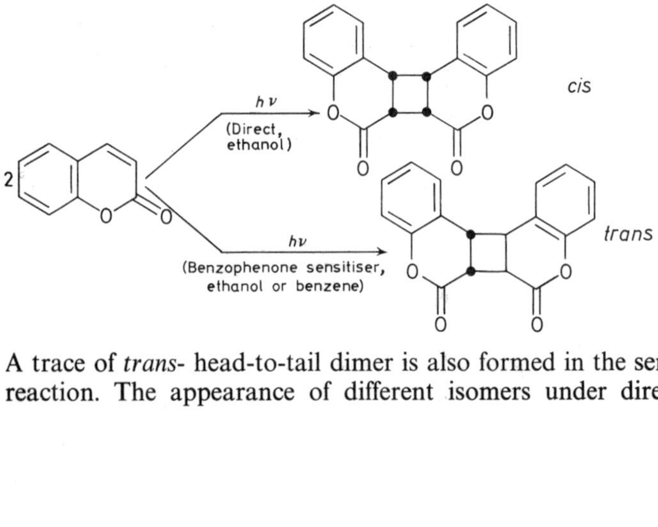

A trace of *trans*- head-to-tail dimer is also formed in the sensitised reaction. The appearance of different isomers under direct and

sensitised irradiation suggests that reaction (6.31a) proceeds via the excited coumarin singlet, and that reaction (6.31b) involves the triplet: in both cases, the molecule with which the excited species reacts is in the ground state. None of the *cis*-dimer appears on direct photolysis of benzene solutions of coumarin, although the sensitised reaction occurs in both solvents. This result is also understandable if two different excited states are involved. The participation of the triplet species in formation of the *trans*-dimer has been clearly shown by experiments in which benzophenone is present in such small concentrations that it absorbs virtually no light. Even under these conditions, irradiation gives relatively high yields of the *trans*-compound: S_1 of coumarin ($E \sim 82$ kcal (343 kJ) mol^{-1}) transfers energy to the lower-lying S_1 state of benzophenone ($E \sim 74$ kcal (310 kJ) mol^{-1}), efficient ISC to T_1 occurs in the benzophenone ($E_T \sim 69$ kcal (289 kJ) mol^{-1}), and triplet excitation is handed back to the coumarin ($E_T \sim 62$ kcal (259 kJ) mol^{-1}). In very dilute solutions of coumarin, and in the absence of benzophenone, some *trans*-dimer is also formed, presumably because at low concentrations unimolecular ISC $S_1 \rightsquigarrow T_1$ can compete with bimolecular reaction of S_1 with S_0.

Excited aldehydes and ketones may take part in reactions which involve the addition of the excited carbonyl group itself to suitable olefins. The product is an oxetane (substituted trimethylene oxide). For example, irradiation of benzophenone in the presence of trimethylethylene leads to the production of an oxetane, and the quantum yield for the process is about 0·5:

$$\text{(6.32)}$$

The triplet (n,π*) state of the carbonyl compound appears to be necessary for the addition, since those 'abnormal' ketones which do not undergo efficient photoreduction (Section 6.5) are also inefficient in oxetane formation. Further, if the *alkene* has a triplet energy below that of the carbonyl compound, energy transfer from the carbonyl group takes place to the virtual exclusion of oxetane formation.

Photochemical addition reactions of oxygen are thought to be important in many photosensitised oxidations of unsaturated compounds. The biological implications of photosensitised oxidation have been recognised since 1900, when it was discovered that micro-organisms can be killed in the presence of oxygen and sensitising dyes. The pathological effects of photo-oxidation of cell constituents include cell damage, induction of mutations or cancer, and death. Recent investigations of photosensitised oxidation have led to a better understanding of the chemical processes, and the results are now finding application in the biological field. It seems appropriate to end the present chapter with a description of these highly significant photo-oxidation reactions.

Almost all photosensitised oxidations proceed via a triplet state of the sensitiser, presumably because this state has a much greater lifetime than the excited singlets. The first process involving the sensitiser triplet may be reaction either with the substrate or with the oxygen. For many dye triplets, the efficiency of reaction with oxygen is so great that this process predominates at all but the lowest oxygen concentrations. Whether the presence of oxygen merely inhibits the reaction with substrate, or whether the products of the primary reaction with O_2 take part in further reactions leading to oxidation of substrate, depends on the chemical nature of the substrate. A very efficient sensitised oxidation is observed with olefins, dienes, dienoid heterocycles and polycyclic aromatic compounds, and it is with these substances as oxidisible reactant that we are concerned. The first oxidation products are peroxides or hydroperoxides, and they may subsequently take part in secondary oxidation steps.

As typical examples, we may consider the products formed on irradiation of dimethylfuran and of tetramethylethylene in the presence of oxygen and a triplet sensitiser; in the first case an endoperoxide (in fact an ozonide) is formed, while the product of the olefin oxidation is a hydroperoxide:

$$H_3C-\underset{O}{\boxed{}}-CH_3 \; + \; O_2 \; \xrightarrow{h\nu,\text{sens}} \; H_3C-\underset{O-O}{\underset{O}{\boxed{}}}-CH_3 \tag{6.33}$$

$$\underset{H_3C}{\overset{H_3C}{>}}C=C\underset{CH_3}{\overset{CH_3}{<}} \; + \; O_2 \; \xrightarrow{h\nu,\text{sens}} \; \underset{H_3C}{\overset{H_3C}{>}}\underset{OOH}{\overset{|}{C}}-C\underset{CH_3}{\overset{CH_2}{<}} \tag{6.34}$$

The kinetic data are consistent with the quantitative formation of some intermediate from the sensitiser triplet at oxygen concentrations above about 10^{-5} mol 1^{-1}. The intermediate is itself trapped quantitatively by good acceptors, although it fails to react at all with many compounds (e.g. alcohols). There are two possible interpretations of these facts: (1) that the intermediate is a sensitiser–oxygen complex, and (2) that the intermediate is electronically excited oxygen formed by energy transfer from sensitiser to oxygen. These two paths probably both involve the formation of an adduct (an 'exiplex') between excited sensitiser and oxygen, but only in the first mechanism is the exiplex stable:

$$\text{(i)} \quad {}^3\text{Sens} + \text{O}_2 \longrightarrow \text{Sens} - \text{O} - \text{O} \tag{6.35}$$

$$\text{(ii)} \quad {}^3\text{Sens} + \text{O}_2 \longrightarrow \text{Sens} + \text{O}_2\dagger \tag{6.36}$$

where ${}^3\text{Sens}$ represents the triplet state of the sensitiser.

Complex formation has, until recently, been regarded as the more probable route to oxidation. However, overwhelming evidence now points to the occurrence of the energy transfer process in many sensitised photo-oxidations. Since the ground state of oxygen is a triplet, it is necessary for $\text{O}_2\dagger$ to be a singlet state for reaction (6.36) to be spin-conserved. The lowest-lying excited state of oxygen is, in fact, a singlet $({}^1\Delta_g)$ and possesses an excitation energy of 22 kcal (92 kJ) mol^{-1}, so that it can easily be excited by energy transfer from the triplet states of most dyes.

Direct studies of the reactions of $\text{O}_2({}^1\Delta_g)$ have indicated that it is the intermediate involved in sensitised photo-oxidation. The excited species may be produced in a number of ways: for example, in the reaction of sodium hypochlorite with hydrogen peroxide (cf. p. 120) or by the action of a microwave discharge on molecular oxygen in the gas phase. For a wide variety of acceptors which yield more than one oxidation product, the product distributions from the reaction with $\text{O}_2({}^1\Delta_g)$ and from the photo-oxidation are identical, and there are no detectable differences in stereoselectivity. If the photo-oxidation involved a bulky sensitiser complex with O_2 at the transition state, a quite different stereoselectivity and product distribution might be expected. Further, the ratio of rate constants for decay and for reaction with acceptor are identical for $\text{O}_2({}^1\Delta_g)$ and for the intermediate in photo-oxidation. It has also been shown explicitly that $\text{O}_2({}^1\Delta_g)$ can be formed in reaction (6.36); the emission band at 1270 nm, from the $\text{O}_2({}^1\Delta_g \to {}^3\Sigma_g^-)$ transition, is observed on irradiation in the gas phase of mixtures of oxygen with suitable triplet donors (e.g. benzaldehyde). This piece of evidence adds considerable weight to the arguments favouring $\text{O}_2({}^1\Delta_g)$ as the intermediate in the sensitised photo-oxidations.

The elucidation of the mechanism of sensitised photo-oxidation has made possible several fruitful speculations with regard to photobiology. As an example, we will consider the protective action of carotenoids in biological systems. Carotenoids apparently protect photosynthetic organisms against the lethal effects of their own chlorophyll (see p. 233), which is an excellent sensitiser of photo-oxidation. The way in which this protection occurs has not hitherto been understood. Recently (1968), however, it has been shown that β-carotene is an extremely efficient quencher of singlet oxygen, and it can also inhibit sensitised photo-oxidations. For example, β-carotene, at a concentration of $10^{-4} \, mol \, l^{-1}$, inhibits 95% of the methylene-blue-sensitised oxidation of 2-methyl-2-pentene ($10^{-1} \, mol \, l^{-1}$); under the conditions of the experiment ($[O_2] = 10^{-2} \, mol \, l^{-1}$), virtually no triplet methylene blue is quenched by the β-carotene, so that the inhibition derives from the quenching of $O_2(^1\Delta_g)$. Carotene does not appear to be consumed in the reaction, which suggests that the quenching of $O_2(^1\Delta_g)$ may involve excitation of triplet carotene by energy transfer:

$$O_2\dagger(^1\Delta_g) + \beta\text{-carotene} \longrightarrow O_2(^3\Sigma_g^-) + {}^3\beta\text{-carotene} \qquad (6.37)$$

Thus the interesting speculation may be made that carotenoids serve a double function in photosynthetic organisms: first, that they remove 'toxic' singlet oxygen, and, secondly, that they can store the energy which O_2 receives from chlorophyll and which would otherwise be lost.

BIBLIOGRAPHY

O. L. CHAPMAN, 'Photochemical rearrangements of organic molecules', *Adv. Photochem.* **1**, 323 (1963)

P. J. WAGNER and G. S. HAMMOND, 'Properties and reactions of organic molecules in their triplet state', *Adv. Photochem.* **5**, 21 (1968)

N. J. TURRO, *Molecular photochemistry:* Chapter 6, 'Photoreduction and related reactions'; Chapter 7, 'Photochemical rearrangements and isomerizations'; Chapter 8, 'Photochemical cycloadditions; Chapter 9, 'Photochemical fragmentations and related reactions'. Benjamin, New York (1966)

D. R. ARNOLD, 'The photocycloaddition of carbonyl compounds to unsaturated systems: the synthesis of oxetanes, *Adv. Photochem.* **6**, 301 (1968)

C. S. FOOTE, 'Mechanisms of photosensitised oxidation', *Science* **162**, 963 (1968)

7

Techniques in photochemistry

7.1 INTRODUCTION

A proper understanding of the principles of photochemistry requires some appreciation of the methods used in the various studies; the present chapter provides a short description of the more common techniques. Detailed discussion of apparatus is, however, purposely excluded; several articles provide this more specific information (see Bibliography at end of chapter).

Photochemical processes may lead to chemical change; the nature of the products, and the rates of their formation, may be determined by standard chemical techniques which need not be treated here. We are more concerned with those parts of the experimental technique which involve light. Measurements of absorbed (and, sometimes, emitted) light intensities are essential to determinations of quantum yields, which are themselves needed in any assessment of the efficiency of primary photochemical processes. Quantum yields are generally determined by the use of 'classical' techniques: that is, under steady illumination. The kinetic behaviour of reaction systems under continuous illumination is often consistent with the presence of reactive intermediates at their stationary state concentrations. Further kinetic data (individual rate constants, for example) may be calculable from a comparison of the system under stationary and non-stationary conditions: this point has already been illustrated with respect to fluorescence decay (p. 94). Photochemical processes are ideally suited to study under non-stationary conditions, since the illumination may be started or stopped suddenly by use of a flash of light or by use of a mechanical shutter: it is often impossible to start or stop thermal reactions in the same way (although shock waves may, of course, be used to cause rapid heating in gaseous systems). In this chapter we first provide a brief summary of classical photochemical techniques, and then describe 'pseudo-stationary' and non-stationary methods.

7.2 CLASSICAL TECHNIQUES

Ideally, a photochemical experiment employs monochromatic light, since the nature of the primary processes, and their quantum efficiencies, may be wavelength-dependent. The use of mono-chromatic radiation also simplifies the measurement of absolute light intensities. Since most light-sources are polychromatic, some technique must therefore be used to isolate a narrow wavelength band. Grating or prism monochromators are well suited to this purpose, although the light intensities available from them may not be sufficient for certain experiments. It is, perhaps, more usual to employ one or more 'colour filters': these may be liquid solutions or glasses of substances which strongly absorb light of unwanted wavelengths. Interference filters, which depend on the interference effects in thin films (akin to those giving rise to colours in soap bubbles), are finding increasing use among photochemists, since they may be constructed to have any desired transmission charac-teristic. Incandescent or gas-discharge electric light-sources are virtually always used. The incandescent tungsten-filament lamp gives radiation whose continuous spectrum approximates to that of a black-body, and, while ordinary lamps are often adequate for the production of visible radiation, extremely high operating tempera-tures are required to obtain significant intensities in the ultra-violet. The introduction of a little iodine into the lamp permits these high filament temperatures to be used without breakdown of the lamp: 'quartz-halogen' lamps (possessing a quartz bulb†) provide a useful, spectrally continuous source of ultra-violet down to $\lambda \sim 200$ nm.

†Ordinary window glass (1 mm thick) transmits about 50% of the incident radiation at $\lambda \sim 320$ nm, while quartz of the same thickness has a transmission of 50% at $\lambda \sim 175$ nm (depending on the purity). All optical components (lamps, windows, lenses, reaction cells, etc.) must obviously be made from materials which are trans-parent at the wavelengths of interest. Approximate wavelengths at which the transmittance of 5 mm thick samples is 50% are, for various substances: window glass, 350 nm; Pyrex, 330 nm; crystal quartz, 185 nm; ultra-pure quartz, 165 nm; CaF_2, 135 nm; LiF, 107 nm. Thinner samples provide useful transmission at significantly shorter wavelengths. Oxygen absorbs at wavelengths less than about 200 nm, and systems for study at $\lambda < 200$ nm must be air-free; at shorter wavelengths it is usual to employ evacuated systems (hence the term 'vacuum ultra-violet' for the spectral region with $\lambda < 200$ nm).

Although there are purposes for which the continuum from incandescent lamps is useful, much higher intensities of near-monochromatic radiation are obtained by filtering radiation from lamps which emit most of their energy in a small number of narrow bands or lines. Several types of gas-discharge lamp may be employed for this purpose; they may contain inert gases or vapours of volatile (often metallic) elements which produce the appropriate atomic emission lines. At low pressures most of the emitted energy may be concentrated in the 'resonance' lines (those due to transitions from the first excited state to the ground state), and essentially monochromatic light can be obtained without the use of filters: typical of such lamps are those containing low-pressure Xe ($\lambda = 147.0$ nm) or Hg ($\lambda = 184.9$ nm, 253.7 nm, cf. p. 37): in the latter case, a little inert gas is usually present, but contributes little to the actual emitted radiation. At higher pressures, created in discharges through metal vapours by operating the lamp at high temperature, more emission lines appear, and they show pressure broadening; the resonance line itself is often *reversed* as a result of absorption by cooler metal vapour near the walls of the lamp. Mercury discharge lamps are very frequently used in photochemical experiments, and Table 7.1 shows the intensities of the main lines from typical low-pressure (intensities relative to $\lambda = 253.7$ nm) and medium-pressure (intensities relative to $\lambda = 365.0$ nm) sources.

Table 7.1. Relative intensities of some lines from mercury discharge lamps. Adapted from J. G. Calvert and J. N. Pitts, Jr., *Photochemistry*, Table 7–2, p. 696, Wiley, New York (1966)

| λ(nm) | Relative intensity (energy units) | |
	Low-pressure	Medium-pressure
579·0 (yellow)	10·1	76·5
546·1 (green)	0·88	93·0
435·8 (blue)	1·00	77·5
404·7 (violet)	0·39	42·2
365·0 (u.v.)	0·54	100·0
334·1	0·03	9·3
313·0	0·60	49·9
303·0	0·06	23·9
296·7	0·20	16·6
289·4	0·04	6·0
265·3	0·05	15·3
253·7	100	reversed

The beam of radiation emerging from the combination of lamp and wavelength-selecting device is now allowed to fall on to the photochemically active reaction mixture. For quantitative investigations, this reaction mixture is usually contained within a cell which has two parallel plane windows normal to the direction of the incident beam. If the beam is itself near-parallel, the light is absorbed evenly over the whole sample. Any light not absorbed emerges from the rear cell window; in a typical experimental arrangement this transmitted radiation may be allowed to fall on some detector used to measure the intensity (see below). Figure 7.1 shows one common optical arrangement for photochemical experiments in the near-ultra-violet: note that the

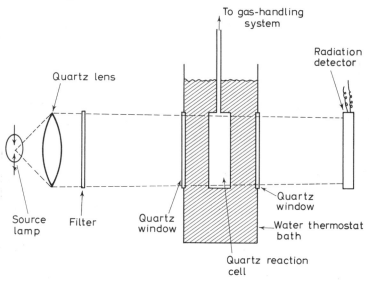

Figure 7.1. One form of apparatus for 'classical' photochemical experiments using near-ultra-violet radiation

components are arranged so that the light beam is almost parallel (perhaps slightly convergent) and so that the beam nearly, but not quite, fills the face of the reaction cell.

We come now to the question of measurement of absorbed light intensity. In principle, we need to know the intensity of the light incident on the front surface of the absorbing substance and also the fraction of light absorbed. The fraction of light absorbed may, in fact, be calculated directly from the measured concentration and a known extinction coefficient for the absorber, by use of the Beer–Lambert law (Section 2.4, p. 25). Alternatively, the detector used for intensity measurements (cf. Fig. 7.1, and next paragraph) can be used to determine *relative* transmitted intensities in the absence of absorber and with the absorbing substance at the desired concentration in the cell: the fractional absorption at the wavelength of the experiment can be calculated directly.

Light is a form of energy and it may be degraded to heat; if light falls on a blackened surface, the temperature of the surface will rise. The temperature rise is commonly measured by a *thermopile*: this is an assembly of thermocouples, connected in series, whose front junctions are blackened (Fig. 7.2). A temperature difference between the illuminated front junctions and the unexposed rear junctions produces an e.m.f., whose magnitude can be measured by a galvanometer or other suitable device. The thermopile may be calibrated against a black-body of known temperature, since the overall energy output of such a source is well known; secondary standards (e.g. standard carbon filament lamps) are, however, often more convenient calibration sources. The energy of the radiation in the photochemical experiment is now calculable, and can be converted into units of quanta s^{-1} via the Planck relation ($E = h\nu$). It is necessary to determine what fraction of the radiation from the black-body, or from the photolysing beam, is actually collected by the thermopile; the lens system of the photochemical experiment is often arranged so that the light-beam converges and is fully accepted by the thermopile. Corrections must also be made for losses such as the reflective losses at the rear windows of the cell and the thermostat tank.

Thermopiles are notoriously sensitive to small fluctuations in room temperature and to draughts, and it is more usual to employ a *photocell* in the chemical experiments. A photocell is shown diagramatically in Fig. 7.2(b); it consists of a photocathode and a collector enclosed in an evacuated bulb. Illumination of suitable cathode materials causes the ejection of electrons, and if the collector is charged positively (i.e. if it is the anode) with respect to the cathode, a current will flow in the external circuit. The operating conditions can be chosen so that this current is proportional to the light intensity reaching the photocathode. However, the quantum efficiency of photoelectron emission from the cathode may not be known, and is, in any case, dependent on wavelength. It is necessary, therefore, to calibrate the photocell against a thermopile, or against a secondary standard. The great advantages of the photocell are, first, that it is more sensitive than the thermopile; and, secondly, that the cathode need not respond significantly to long-wavelength radiation, so that small temperature fluctuations are no longer tiresome. Indeed, the cathode material for ultra-violet intensity measurements can be chosen (e.g. pure

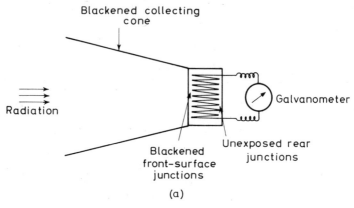

(a)

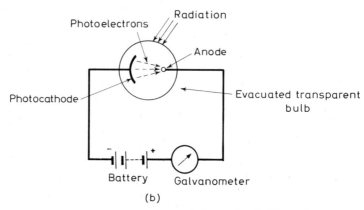

(b)

Figure 7.2. Radiation detectors: (a) thermopile; (b) photocell

sodium) so that the photocell does not detect visible light, and stray illumination from the laboratory lighting need not then be rigorously excluded.

An alternative approach to the determination of light intensities is to measure the rate of a photochemical reaction *for which the quantum yield is accurately known.*† Chemical systems of this kind are referred to as *chemical actinometers*. Bunsen and Roscoe, in the last century, first used a chemical actinometer to measure light intensities. Figure 7.3 shows schematically the apparatus used. The reaction was the $H_2 + Cl_2$ reaction, and the product HCl dissolved in water, so that the water moved along the capillary: the rate of movement is an indicator of the light intensity. Unfortunately, the

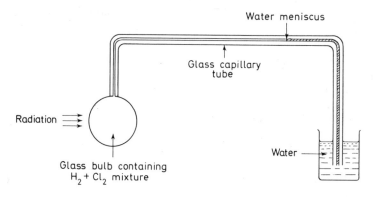

Figure 7.3. The Bunsen–Roscoe chemical actinometer

H_2/Cl_2 actinometer possesses several qualities most undesirable in an actinometer. The quantum yield is unpredictable, wavelength-dependent, and highly sensitive to experimental conditions (e.g. pressure, temperature, intensity, presence of trace impurities, size and material of reaction vessel). Chemical actinometers are chosen today for their insensitivity to wavelength and the experimental parameters. Until recently, uranyl oxalate solutions have been the actinometers of choice in most applications (cf. p. 82). However, the use of low fractional conversions means that there is a relatively small change in the oxalate concentration, and the difference in permanganate titres for oxalate may be small and subject to error. Potassium ferrioxalate has now replaced uranyl oxalate as the standard substance for chemical actinometry. Irradiation of ferrioxalate results in reduction of Fe^{3+} to Fe^{2+} (and simultaneous oxidation of the oxalate ion). In the usual procedure (Hatchard and Parker, *Proc. Roy. Soc. (London), Ser. A,*

---

†This quantum yield itself has, of course, to be measured by thermopile determinations of light intensity.

235, 518 (1956)), the ferrous ion produced is estimated absorptiometrically after formation of the red complex with o-phenanthroline. Since no ferrous iron is initially present, and since the absorption of the red complex is intense, it is easy to demonstrate the formation of, say, 10^{-8} mol of Fe^{2+}. The quantum yield for Fe^{2+} formation is substantially constant over the wavelength range 254 nm–579 nm, and is virtually insensitive to temperature, solution composition and light intensity.

The most widely used gas-phase actinometer is acetone. For the wavelength region 250 nm–300 nm, and with temperatures above 125°C and pressures below 50 torr, the quantum yield for CO formation is unity. The liquid phase ferrioxalate actinometer is, however, more generally useful for measurement of light intensities.

A chemical actinometer may either be used in a second experiment in place of the substance on which the photochemical experiment is being performed, or be placed in a second cell behind the main reaction cell. In the latter case, rates of reaction in the actinometer with and without the absorber in the first, main, cell can be used to calculate the intensity of light absorbed in the first cell. In either case, suitable corrections must be applied for reflections at the various interfaces; further, if a liquid actinometer is used for intensity measurements to be applied to gas-phase photochemical reactions, allowance must be made for the differences in gas–cell and liquid–cell interface reflectivity. The most satisfactory way of using a chemical actinometer may, in fact, be to calibrate a photocell against it, so that in subsequent experiments a given photocell current corresponds to a known light intensity incident on the reaction mixture.

7.3 EMISSION STUDIES

The quantitative study of luminescence requires the use of certain special techniques, some of which are described in this section. Intensities of fluorescence, phosphorescence and chemiluminescence are generally much smaller than those of the irradiating light used for photolysis or excitation, and sensitive methods of detection are needed. Photographic records of emission spectra can yield intensity data averaged over the exposure period as well as providing information about the spectral distribution of the emission. Photoelectric methods are, however, normally used in quantitative investigations because of their superior sensitivity and speed of response. Photocells, such as those described in the last section, can be made to detect radiation of wavelengths as long as about 1300 nm by choice of a suitable cathode (Ag–O–Cs); the short-wavelength limit is set more by the transmission of the cell windows than by the response of the photocathode, and it may be most convenient to coat the front of the cell with a fluorescent material which converts ultra-violet radiation to visible emission detectable with a glass-envelope photocell. Small photocell currents may be amplified by conventional electronic techniques, and low-intensity emission can be detected in this way. Some noise is inevitably introduced by such amplification, and weak emission is best detected with *photomultipliers*. A photomultiplier is effectively a photocell possessing internal amplification which is almost noise-free. Figure 7.4 shows diagramatically the construction of a photomultiplier and the external circuitry employed. A photo-electron ejected from the cathode is accelerated by the electric field towards the first *dynode*; the kinetic energy of impact of the photoelectron on the dynode is sufficient to eject many secondary electrons from the surface material. Each of these secondary electrons is accelerated to the second dynode, each ejects secondary electrons from this dynode, and so on. Current (i.e. numbers of electrons per second) gains of $10^6–10^7$ are easily obtained with modern photomultipliers, and since, in principle, only electrons liberated photoelectrically from the cathode can initiate the liberation of secondary electrons at the dynodes, no noise is introduced by the current-amplifying dynode chain. There is, of course, a small number of electrons liberated thermally from the cathode material (and, to an even lesser extent, from the dynodes); cooling the photomultiplier (e.g. in liquid nitrogen) reduces the small thermal 'dark current', and cooled photomultipliers are used for the detection of exceedingly low intensity radiation.

The spectroscopic nature of luminescent emission may be investigated by the use of a dispersing instrument (i.e. mono-chromator) together with a photomultiplier to detect radiation. The spectroscopic response of the photomultiplier must be known in order to determine the true emission spectrum: the spectral sensitivity of photomultipliers is discussed later in connection with absolute calibrations. Fluorescence or phosphorescence *excitation spectra* are obtained by monitoring the emission intensity (preferably within a narrow wavelength band) as the wavelength of the exciting light is altered: the true excitation spectrum is obtained only if the intensity of exciting light is constant at all wavelengths,

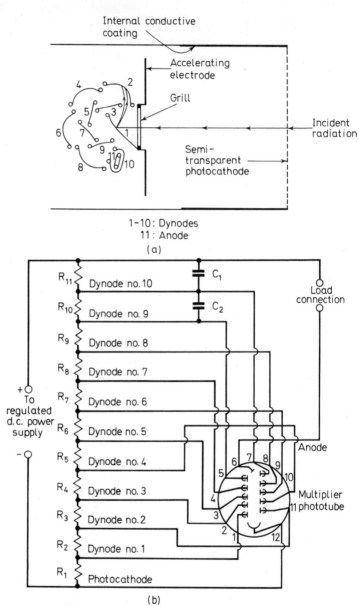

Internal conductive coating

Accelerating electrode

Grill

Incident radiation

Semi-transparent photocathode

1–10: Dynodes
11: Anode

(a)

R_{11} Dynode no. 10 C_1 Load connection

R_{10} Dynode no. 9 C_2

R_9 Dynode no. 8

R_8 Dynode no. 7

R_7 Dynode no. 6

R_6 Dynode no. 5

Anode

R_5 Dynode no. 4

R_4 Dynode no. 3

R_3 Dynode no. 2

R_2 Dynode no. 1

R_1 Photocathode

+ To regulated d.c. power supply

−

Multiplier phototube

(b)

Figure 7.4. The photomultiplier: (a) Construction of one type of photo-multiplier; (b) external electrical circuit

and if the intensity is not constant, appropriate corrections must be applied to the observed excitation spectrum.

Fluorescence and phosphorescence are almost always observed at right angles to the direction of the irradiating beam. Scattered light of the incident radiation can be troublesome, especially if the

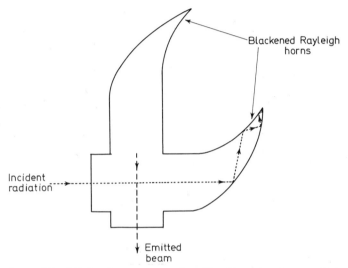

Figure 7.5. Cell for fluorescence studies. The illustration shows multiple reflections of the incident beam into the blackened Rayleigh horn

luminescence is weak, and the presence of dust or polycrystalline solids often precludes the study of luminescence lying at wavelengths near that of the exciting light. Reflections of the exciting light by the cell faces can be virtually eliminated by suitable cell design: Fig. 7.5 shows the use of 'Rayleigh Horns' in a fluorescence cell. If an irradiated system exhibits both fluorescence and phosphorescence, it may be difficult to establish the contribution to emission made by each process. The basic technique used to differentiate between the two phenomena was devised by Becquerel in 1859; present-day methods are based on modifications of an apparatus proposed by G. N. Lewis: Fig. 7.6 shows a plan view of one form of the apparatus. The sample is enclosed within a rotating drum which has a slot cut in it parallel to the axis of rotation. As usual, the directions of irradiation and observation are perpendicular, but, because of the single slot, the observations are made only after a time lag following irradiation. The delay is determined by the speed of rotation, and can be set so that fluorescence has decayed almost entirely by the time of observation while phosphorescent emission is virtually unaffected: with a motor rotating the drum at, say, 6000 rev/min the delay time is 2·5 ms, which lies conveniently between typical radiative lifetimes of fluorescent and phosphorescent emissions.

Radiative lifetimes of emission phenomena may be measured by use of the phosphorimeter, since the intensity of *observed* emission

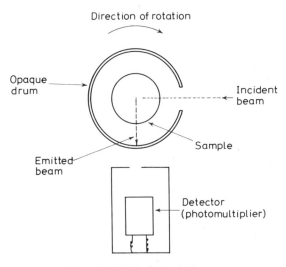

Figure 7.6. Plan of phosphorimeter

can be measured as a function of speed of cylinder rotation: that is, as a function of time after irradiation ceases. The decay of emission intensity is usually kinetically first-order, and it is not necessary to measure absolute emission intensities. (The integrated form of the first-order equation is

$$\ln I_0 - \ln I_t = t/\tau \tag{7.1}$$

where I_t is the intensity at time t, I_0 the intensity at $t = 0$ and τ the radiative lifetime. A plot of $\ln I_t$ against t will therefore have a slope $1/\tau$ no matter what units are used for I_t: the appropriate scaling factors will be taken up in the intercept.) The speed of drum rotation places a limit on the shortest lifetime that can be measured by this technique: lifetimes greater than about 10^{-6} s can be measured fairly readily.

Different techniques must be adopted for the measurement of fluorescence lifetimes (i.e. with $\tau < 10^{-7}$ s). One method involves the use of a discharge lamp which produces a flash of extremely short duration ($\sim 10^{-9}$ s) for excitation.† The fluorescent emission is detected by a photomultiplier whose output is displayed on the vertical axis of an oscilloscope; the oscilloscope trace (horizontal axis $\equiv$ time) is triggered by the flash, and the first-order decay of fluorescence with time is shown directly on the screen. As before, the radiative lifetime may be calculated from relative intensities. A correction can be made for the finite duration of the exciting flash (the intensity–time profile can be established by monitoring the exciting intensity directly, and displaying the photomultiplier output on the oscilloscope), and fluorescence lifetimes in the nanosecond range can be measured.

An alternative method for the determination of fluorescence lifetimes uses modulated light for excitation of the emission. There is an average delay between the absorption of a photon and emission of fluorescence, which corresponds to the lifetime of the excited state. Thus, although the fluorescent emission is modulated at the same frequency as the exciting radiation, there is a phase difference between the two modulations. The phase angle, ϕ, may be measured by electronic comparison of the modulation present on the exciting and fluorescent beams; it is related to the mean fluorescence lifetime, τ, by the expression

$$\tan \phi = 2\pi v \tau \qquad (7.2)$$

where v is the modulation frequency. Fluorescence lifetimes may be measured, by this technique, with an accuracy approaching half a nanosecond.

Absolute quantum efficiencies for fluorescence or phosphorescence can be calculated from measurements in the *same arbitrary units* of absorbed and emitted intensities. Allowances must be made for the differences in spatial and spectral distributions of exciting and emitted radiation, and the spectral response curve of the detector must be known. The directed exciting beam may be scattered

---

†This method belongs to the class of non-stationary techniques described in Section 7.5, but is included here as the application is straightforward. Properly speaking, the phosphorimeter technique is a 'pseudo-stationary' technique (Section 7.4).

for comparison with isotropic emitted radiation either by a matt surface, or, better, by a protein solution of calculable scattering power. A simplification of the corrections required for the spectral distribution of emitted light may be achieved by allowing the emitted radiation and the scattered exciting beam to fall successively on a suitable fluorescent substance which converts all incident radiation *to its own fluorescence spectrum with uniform quantum efficiency*. Such a device is called a 'quantum counter', and it is clearly necessary that the fluorescence quantum yield be wavelength-independent for all wavelengths in the emitted and exciting radiation; it is *not*, however, necessary to know the absolute value of the quantum yield for the counter material. One substance frequently employed is a solution of rhodamine B. The use of the quantum counter removes problems concerned both with the spectral sensitivity of the detector and with the spectral distributions of exciting and emitted light, since the detector always receives the same fluorescence spectrum of the counter material regardless of the excitation wavelength. It may be noted that the relative spectral sensitivity curve of a detector can be determined by comparison of response, at a series of wavelengths, to a monochromatic scattered beam viewed directly, and to the fluorescent emission excited by the radiation in a quantum counter.

Direct evaluation of quantum yields of emission processes by measurement of *absolute* emission (and absorption) intensities is possible, although the low intensity of many emission processes makes such measurements difficult. The absolute intensities may be determined either by the primary thermopile standard, or by a previously calibrated photomultiplier. The potassium ferrioxalate chemical actinometer may also be employed for absolute emission intensity measurements because of its high sensitivity.

The absolute quantum yield of a luminescent process may, of course, be assessed by comparison of emitted intensities from the sample and from a substance whose emission quantum yield is already accurately known. Although this method begs the question of how the quantum yield was originally determined for the standard substance, it may in practice be the most convenient and rapid technique. Several standards have been suggested. One useful substance is the sodium salt of 1-naphthylamine-4-sulphonic acid; dilute, oxygen-free, solutions in glycol exhibit a quantum yield (determined by one of the 'absolute' methods) close to unity. A similar technique is finding increasing application in measurements of absolute intensities in gas-phase chemiluminescence. The spectral distribution and absolute emission efficiency of the 'air-afterglow' chemiluminescence (p. 123)

$$O + NO + M \longrightarrow NO_2\dagger + M \tag{7.3}$$

is known from experiments using the potassium ferrioxalate actinometer. It is a simple matter to measure the intensity of the air afterglow resulting from the reaction of known concentrations of O and NO and compare it with the intensity at any wavelength of a process whose efficiency is to be measured. Since the geometrical conditions are identical for 'known' and 'unknown' chemiluminescence, it is unnecessary to estimate the fraction of the total emitted light which is received by the detector.

7.4   PSEUDO-STATIONARY METHODS

In order to illustrate the kinetic applications of experiments performed under non-stationary conditions, we shall consider first the kinetics of a simple reaction, and see how they are affected by repeated interruption of the photolysing beam (it is this repetition of a non-stationary experiment that gives rise to the term 'pseudo-stationary').

At high temperatures we may represent the steps in acetaldehyde photolysis by the equations:

$$CH_3CHO + h\nu \xrightarrow[CH_3CHO]{\phi} 2CH_3 + H_2 + 2CO \tag{7.4}$$

$$CH_3 + CH_3CHO \xrightarrow{k_1} CH_4 + CH_3 + CO \tag{7.5}$$

$$CH_3 + CH_3 \xrightarrow{k_2} C_2H_6 \tag{7.6}$$

The differential equation governing the formation of methyl radicals is

$$\frac{d[CH_3]}{dt} = 2\phi I_{abs} - k_2[CH_3]^2 \tag{7.7}$$

The steady state approximation puts $d[CH_3]/dt = O$, so that

$$[CH_3]_{ss} = \sqrt{\frac{2\phi I_{abs}}{k_2}} \tag{7.8}$$

where the subscript indicates that the concentration is that at the steady state. This value for $[CH_3]$ can be substituted into the appropriate rate equations for methane and ethane formation

$$\frac{d[CH_4]_{ss}}{dt} = k_1[CH_3]_{ss}[CH_3CHO] = k_1\sqrt{\frac{2\phi I_{abs}}{k_2}} \cdot [CH_3CHO]$$

$$(7.9)$$

$$\frac{d[C_2H_6]_{ss}}{dt} = k_2[CH_3]_{ss}^2 = 2\phi I_{abs} \tag{7.10}$$

Measurement of the steady state rate of ethane formation gives a value for $2\phi I_{abs}$ (and if I_{abs} is measured absolutely, the primary quantum yield ϕ can be determined), but a value for k_2 cannot be calculated; even if $d[CH_4]/dt$ is also measured, it is possible only to evaluate the ratio $k_1 k_2^{-\frac{1}{2}}$. Two alternative methods can be used to determine the individual rate constants. First, if the steady state concentration of methyl radicals can be measured directly, then k_1 and k_2 are immediately calculable from the rate data via the first equalities of Eqs. (7.9) and (7.10). Concentrations of reactive intermediates present in chemical reactions can frequently be measured with considerable precision: the physical techniques include, for appropriate radical species, optical spectroscopy, EPR spectroscopy, mass spectrometry and calorimetry. Discussion of these techniques is not possible in this book: a source of further information is suggested in the Bibliography.

The second method consists of making measurements under non-stationary conditions, and it is with this technique that we are now concerned. Let us suppose that the photolysing beam, brought to a focus, can be interrupted by an opaque rotating disc in which a number of sectors have been cut away (see Fig. 7.7); we shall further suppose, for the purposes of illustration, that the opaque sectors remaining and those cut away are of equal size, so that, as the wheel rotates, the reaction mixture is successively irradiated and in darkness for equal periods. The behaviour of the reaction system under the intermittent photolysis depends on the rate at which the light-beam is 'chopped'. Qualitatively, we may express this behaviour in the following way. At very slow chopping speeds the reaction mixture is effectively exposed to the full light intensity for half the total time of the experiment, because the system reaches its steady state rapidly in comparison with the irradiation period. The apparent rate of reaction (e.g. the rate of methane formation) is therefore one-half of that observed in the absence of the sector. However, if the periods during which the light-beam is 'on' and 'off' are short compared with the time taken for the system to reach a steady state, the reaction mixture is effectively exposed for the whole of the time to one-half of the total light intensity: that is, the situation is equivalent to one in which a neutral density filter transmitting 50% of the incident radiation is interposed in

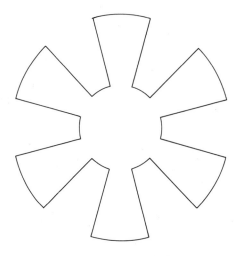

Figure 7.7. Sector wheel

the light-beam in a static experiment. Equation (7.9) shows that the rate of methane formation is proportional to the *square root* of the light intensity, so that with the sector rotating rapidly $d[CH_4]/dt$ is equal to $\sqrt{2}$ of its magnitude in the absence of the sector. The difference between 'slow' and 'fast' sector speeds corresponds to the time taken for the rate of reaction to reach its steady state in comparison with the periods of exposure to radiation. Figure 7.8(a) shows how $[CH_3]$ might vary with time for 'slow' and 'fast' chopping speeds. With rapid sector rotation $[CH_3]$ never approaches the steady state value during the 'on' period, nor does it drop away to zero during the 'off' period. Thus $[CH_3]$ depends on whether the lifetime of the radical, τ, is long or short compared with the interruption period. Figure 7.8(b) indicates the dependence of $d[CH_4]/dt$ on sector speed: the change-over in rate of reaction corresponds to a sector speed which gives light and dark periods of about the magnitude of τ. The average lifetime of a

CH_3 radical under steady state conditions is equivalent to the steady state radical concentration divided by the rate at which radicals are removed:

$$\tau_{ss} = \frac{[CH_3]_{ss}}{-d[CH_3]/dt} = \frac{[CH_3]_{ss}}{k_2[CH_3]_{ss}^2} = \sqrt{\frac{1}{2\phi I_{abs} \cdot k_2}} \qquad (7.11)$$

Hence, from Eqs. (7.10) and (7.11),

$$k_2 = (\tau_{ss}^2 \cdot d[C_2H_6]_{ss}/dt)^{-1} \qquad (7.12)$$

and if $\tau_{ss} \approx \tau$, then k_2 can be calculated.†

At this point it is appropriate to mention another photo-chemical technique for determination of individual rate constants

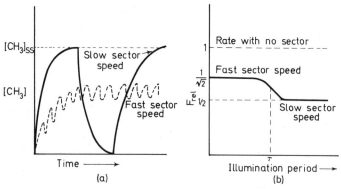

Figure 7.8. The effects of 'slow' and 'fast' sector rotation speeds (a) on the methyl radical concentration and (b) on the rate of methane formation (F_{rel}) relative to the steady state rate in the absence of the sector

which, although bearing little experimental resemblance to the rotating sector method, does, in fact, have a similar basis. This technique is the *spatial interference* method, and is normally applied

†The treatment given indicates the nature of the rotating sector technique. In practice, the value of τ is not read off Fig. 7.8(b), as considerable error could result. Rather, the mean radical concentration for any chopping speed is calculated from the solutions to the differential rate equations for $d[CH_3]/dt$ during light and dark periods (i.e. from Eq. 7.7 during irradiation, and from the second-order decay equation

$$d[CH_3]/dt = -k_2[CH_3]^2 \qquad (7.13)$$

during the dark period). The expressed variation of reaction rate with sector speed is then calculated for various magnitudes of τ, and the value of τ giving the 'best fit' to the *whole* experimental curve of Fig. 7.8 (b) is chosen.

to liquid phase experiments. Let us consider two beams of light, of equal intensity, falling on a photochemically active reaction mixture. If the beams fall on isolated parts of the mixture, the over-all rate of reaction will be twice that found with a single beam. On the other hand, if the two beams are coincident, the reaction mix-ture is effectively exposed to twice the intensity of the single beam. If, as is often the case, the rate of reaction is proportional to the square root of the light intensity (cf. Eq. 7.9: the $\sqrt{I_{abs}}$ term is a consequence of second-order removal of the photochemically generated intermediate), then the rate will be only $\sqrt{2}$ larger than with the single beam. Whether or not the beams may be said to be 'far apart' or 'coincident' depends on whether the radical can diffuse in its lifetime from one irradiated area to the other. Thus a determination of the beam separation at which the rate decreases from twice to $\sqrt{2}$ of the single beam value can be used to calculate the average diffusion time and, hence, the mean radical lifetime, τ_s. It should be noted that the method actually yields a rate con-stant relative to the diffusion rate, and is not 'absolute' in terms of time; in the rotating-sector technique time is measured directly. The practical arrangement usually employs several 'pairs' of narrow beams to give good spatial resolution at the same time as reason-ably fast rates of reaction. Several optical arrangements have been devised: the nature of the most common is self-evident, since the spatial interference method is often described colloquially as the 'leopard spot' technique.

7.5 NON-STATIONARY METHODS: FLASH PHOTOLYSIS

We have seen in the last section how a knowledge of the mean lifetime, τ, of a reaction intermediate may be used in the calculation of individual rate constants for elementary processes. The pseudo-stationary technique provides one possible, and sometimes con-venient, method for the determination of τ, although it is only a special modification of the more general non-stationary technique. If it is possible to follow continuously the rate of reaction, or inter-mediate concentration, as a function of time during the non-stationary phase of a reaction, τ may be evaluated and the individual rate constants may then be calculated. An obvious application of the technique is in the evaluation of kinetic parameters for emission processes. The intensity of emission is proportional to the concentration of excited species, and since the decay is first-order, τ may be determined by the non-stationary technique described on p. 205, without prior knowledge of the Einstein A factor (see also Section 4.2, p. 94 for a discussion of emission lifetimes).

The use of a flash of light to induce photochemical reaction forms the basis of *flash photolysis*. The method was developed initially by Norrish and Porter in an attempt to identify unequivocally the reactive intermediates in a photochemical system. The stationary concentrations of atoms, radicals or excited species present in a static system are normally too low for the intermediates to be detected by their absorption spectra. However, if an extremely high intensity flash source is used, then the transient concentrations of intermediates may be sufficiently large for spectroscopic

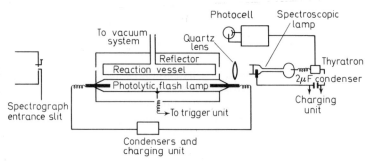

Figure 7.9. Schematic diagram of flash photolysis apparatus. (From R. P. Wayne in Comprehensive Chemical Kinetics *(Ed. C. H. Bamford, and C. F. H. Tipper), Vol. I, Fig. 3, p. 284, Elsevier, Amsterdam (1969))*

observation. Further, the changes with respect to time of the intermediate concentration may be followed by means of the absorption spectrum, and kinetic data, such as radical lifetimes, may be obtained. This use of time-resolved spectroscopy is often known as *kinetic spectroscopy*. (Kinetic spectroscopy may also be used to monitor continuously the concentrations of suitable reactants and final products as a function of time after the flash.) The mechanisms of many photochemical reactions have finally been elucidated with the help of the information given by flash photolysis experiments about the nature and reactivity of intermediates, and since the technique has proved so successful, a short description will be given of the experimental method.

Figure 7.9 shows one form of the apparatus. A flash lamp,

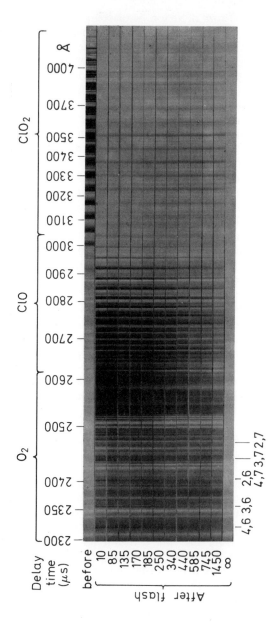

Plate I. *Flash photolysis of chlorine dioxide: spectra of* ClO_2 *and its photolysis products* ($ClO_2 = 1$ *torr*; $N_2 = 200$ *torr*)
(From F. J. Lipscomb, R. G. W. Norrish and B. A. Thrush, Proc. R. Soc., **A233,** *457 (1956))*

containing about 100 torr of xenon, is placed parallel to a quartz reaction cell; both cell and lamp are surrounded by a reflecting enclosure. The lamp is connected to a bank of condensers. The condensers are charged to a high voltage which is, however, insufficient to initiate a discharge in the lamp: the discharge is started by means of a high voltage applied to the trigger electrode. Modern flash lamps are capable of dissipating high powers in short-duration pulses. For example, flashes of 7500 J ($C = 275\mu F$ at $V = 5000$ V) have been obtained with durations of 17 μs, equivalent to a power of nearly 500 MW – about a million times more than the power of a large conventional continuous-running discharge lamp.

A small amount of light from the photolysis flash is allowed to fall on to a photocell, which is connected to an electronic delay unit. The delay unit then triggers a second, low-power, 'spectroscopic' flash lamp, the light from which is directed down the length of the reaction cell and ultimately on to the slit of a spectrograph. A photographic record is thus made of the absorption spectrum of the reaction vessel contents at a fixed time (determined by the electronic delay) after the photolysis flash. The reaction vessel is then refilled with fresh reactant and the experiment is repeated with a different delay time. In this way a series of spectra, corresponding to the different delay times, is obtained. Plate I shows the photographic records obtained in some experiments on the flash photolysis of chlorine dioxide. Absorption bands due to ClO appear after the flash, although they are not present before photolysis or after long delays; a second transient absorption due to vibrationally excited O_2 is also seen, and the spectroscopic results are consistent with the photolysis scheme

$$ClO_2 + h\nu \longrightarrow ClO + O \qquad (7.14)$$

$$O + ClO_2 \longrightarrow ClO + O_2^* \qquad (7.15)$$

$$ClO + ClO \longrightarrow Cl_2 + O_2 \qquad (7.16)$$

It is also possible to make a reliable estimate for the extinction coefficient of the ClO radical and hence, from quantitative plate photometry, to measure the rate constant for the second-order reaction (7.16). Note that where the loss processes for an intermediate are described by a first-order law, it is not necessary to know the extinction coefficient, since measurement of *relative* concentrations permits evaluation of the first-order rate constant.

Many types of transient species have now been studied by kinetic spectroscopy. Flash photolysis of ClO_2, described in the last paragraph, indicates the presence of excited species (vibrationally hot O_2) as well as a radical (ClO) in the photolysis. Atoms and electronically excited species have also been shown to be present in specific photochemical systems by the flash photolysis technique. For example, triplet–triplet absorption spectra following flash photolysis of suitable phosphorescent substances have provided valuable data about the rates of intersystem crossing and triplet quenching: much of this information is implied in the discussions of Chapters 4 and 5. Although photographic records of absorption spectra, obtained in the manner described, are usually employed for the identification of photochemically generated intermediates, accurate quantitative kinetic studies of a specific species are better carried out by monitoring the intensity of a single band or line at a fixed wavelength. The use of a monochromator–photomultiplier combination allows the intensity to be followed continuously with time, and errors introduced in plate photometry, or by the photographic process itself, are eliminated. Photoelectric observations can be made both for emission from excited species (as in the fluorescence lifetime measurements described on p. 205) or for absorption if the spectroscopic flash to be used in the photographic experiments (Figure 7.9) is replaced by a constant-intensity source lamp.

BIBLIOGRAPHY

J. G. CALVERT and J. N. PITTS, JR., *Photochemistry:* Chapter 7, 'Experimental methods in photochemistry', John Wiley, New York (1966)

H. W. MELVILLE and B. G. GOWENLOCK, *Experimental methods in gas reactions:* Chapter 6, 'Photochemical techniques', Macmillan, London (1964)

R. G. W. NORRISH, and B. A. THRUSH, 'Flash photolysis and kinetic spectroscopy', *Q. Rev. chem. Soc.* **10**, 149 (1956)

R. P. WAYNE, 'The detection and estimation of intermediates', in *Comprehensive Chemical Kinetics*, Vol. II (Eds. C. H. BAMFORD, and C. F. H. TIPPER), Elsevier, Amsterdam (1969)

C. A. PARKER, *Photoluminescence of solutions,* Elsevier, Amsterdam (1969)

8

Photochemistry in action

8.1 APPLIED PHOTOCHEMISTRY

The title to this chapter is deliberately vague, and it requires explanation. Earlier chapters of the book have been concerned with the fundamental processes of photochemistry. In contrast, we wish now to examine certain phenomena in which photochemical processes play an important part. We shall consider, therefore, some applications of photochemistry (e.g. photography), and the author hopes that the limited number of examples selected will show the diversity of such applications. *Natural* photochemical phenomena have led to the evolution of life as we know it, and permit of its continued existence on earth, and we shall include in the scope of our title some discussion of these topics.

8.2 ATMOSPHERIC PHOTOCHEMISTRY

Photochemical reactions have performed a determining role in the evolution of the atmosphere and of life on Earth. The understanding of primary photochemical processes that has been reached over the last two decades has permitted reasonable speculation about the history of the atmosphere; the 'investigation' of the Earth's palaeoatmosphere (fossil atmosphere) has in turn suggested solutions to several 'puzzles' concerning the Earth's geology. The forms and ecology of life which were viable at any time in the past were directly dependent on the constitution of the atmosphere at that period; we shall see that, conversely, processes involving living organisms exert a major influence on atmospheric composition. It is this interrelation between atmospheric and biological evolution that makes the study of the Earth's palaeoatmosphere, and comparison with the present-day atmospheres of other planets, particularly rewarding. In this section we put forward one view of the development of the Earth's atmosphere.

Much evidence shows that Earth was without a primordial atmosphere. For example, the abundances of the rare gases in the contemporary atmosphere lie between 10^{-10} and 10^{-6} of their cosmic abundances. It has been shown that the quantities of gases liberated as a result of volcanic activity, and from slow decay of solid radioactive elements, are sufficient to account for our atmosphere. However, *oxygen is not released from volcanic effluents*, and the primitive atmosphere must have contained H_2, N_2, CO_2 and H_2O as its most important constituents.

In the absence of life, the main source of O_2 must have been photolysis of water by short-wavelength ultra-violet:

$$H_2O + h\nu \longrightarrow 2H + O \tag{8.1}$$

$$O + O(+M) \longrightarrow O_2(+M) \tag{8.2}$$

The production of O_2 by this mechanism is limited by the 'shadowing' effect of the O_2 formed. Molecular oxygen is likely to have been distributed in the atmosphere above the H_2O vapour, and to have absorbed the radiation responsible for photodissociation of water (say at $\lambda < 195$ nm). This argument can be extended quantitatively to show that the upper limit for oxygen concentration in the primitive atmosphere was less than 10^{-3} of the present atmospheric level (P.A.L.). Absorption of ultra-violet radiation by CO_2 would add to the shadowing and reduce further the limiting O_2 concentration. Geological evidence is consistent with $[O_2] < 10^{-3}$ P.A.L.: the incomplete oxidation of early sedimentary materials suggests sedimentation in a reducing atmosphere, and such oxides as were formed could have resulted from oxidation by relatively small amounts of *ozone* formed close to the Earth's surface by the three-body process:

$$O + O_2(+M) \longrightarrow O_3(+M) \tag{8.3}$$

(See below for a discussion of present-day ozone distributions.)

Organic substances such as amino acids are rapidly degraded in an oxygenic atmosphere, especially in the presence of sunlight, and low atmospheric oxygen concentrations seem, in fact, to be necessary for the development of the organic precursors of living matter.

Photosynthesis is the only process which can have caused the rise of $[O_2]$ from 10^{-3} P.A.L. to 1 P.A.L. We shall discuss photosynthesis further in Section 8.4, and for the time being we need only note that the process involves consumption of carbon dioxide and water and the concomitant *liberation of oxygen*. It can be shown that, in the present atmosphere, all O_2 passes through the photosynthetic process in the extremely short period (by geological standards) of about 2000 years, and photosynthesis is clearly an efficient source of O_2. The build-up of oxygen in the atmosphere is dependent on a rate of O_2 production (mainly a result of photosynthesis) in excess of the rate of loss (resulting from oxidation,

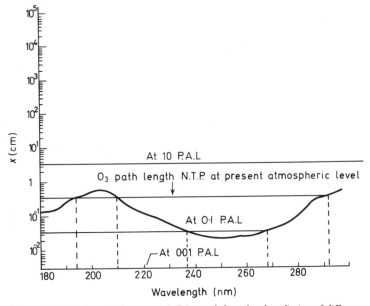

Figure 8.1. Path-length of ozone at S.T.P. needed to absorb radiation of different wavelengths so that the transmitted intensity is 1 erg cm^{-2} s^{-1} in a 5 nm band. The horizontal lines represent equivalent O_3 path-lengths calculated for three oxygen concentrations. (From L. V. Berkner and L. C. Marshall, Discuss. Faraday Soc. **37,** *122, Fig. 10 (1964))*

respiration, etc.). In the first stages of progress from $[O_2] \sim 10^{-3}$ P.A.L., photosynthetic activity (at about present densities) must have covered a few percent of the (present) continental areas before there was a net positive balance contributing to atmospheric oxygen.

The cells responsible for photosynthesis are sensitive to ultraviolet radiation, and wavelengths shorter than about 300 nm are lethal to most living organisms. Nucleic acids absorb in the wavelength region 260–270 nm, and proteins from 270 to 290 nm, and ozone is the only constituent of the atmosphere which filters these wavelengths from solar radiation. Figure 8.1 shows the thickness of O_3, at STP, required to absorb radiation so that the transmitted intensity in a 5 nm band is 1 erg $(10^{-7} J)$ cm^{-2} s^{-1} (in the visible region, this energy corresponds to about 50 times the brightness of the full moon). Atmospheric ozone concentrations may be calculated from the ordinary photochemical scheme (see p. 223) for any oxygen concentration, and the horizontal lines in Fig. 8.1 show these concentrations, expressed in equivalent path lengths at STP, for $[O_2] = 10^{-2}$ P.A.L., 10^{-1} P.A.L. and 1 P.A.L. The important conclusion is that at $[O_2]$ below 10^{-1} P.A.L. there would be *insufficient* ozone to protect organisms from lethal radiation in the wavelength range 260–300 nm. It appears to follow that life must have existed beneath liquid water, which in sufficient thickness can screen ultra-violet radiation, until $[O_2]$ reached 10^{-1} P.A.L. At $[O_2] = 10^{-3}$ P.A.L. the depth of water needed to filter lethal radiation is about 10 m, and primitive organisms probably lived in shallow lakes; the 'most satisfactory' depth would represent a compromise between shielding of ultra-violet radiation and transmission of the visible light required for photosynthesis. Life in the oceans seems improbable at this stage, since organisms would be brought too near the surface by mechanical motions.

The oceans can support life when the penetration of lethal radiation is reduced to a few centimetres, a situation arising when $[O_3]$ increases to a level corresponding to $[O_2] \sim 10^{-2}$ P.A.L. At the same time the activity of many primitive organisms changes as $[O_2] \rightarrow 10^{-2}$ P.A.L. Fermentation gives way to respiration, and photoreduction to photosynthesis. Thus one might anticipate a sudden increase in evolutionary activity when this oxygen concentration is achieved. For many years the absence of fossil remains from periods earlier than the Cambrian (600 million years ago) has been regarded as a geologic 'puzzle'. It now appears that the evolutionary explosion at the beginning of the Cambrian period is to be identified with atmospheric oxygen concentrations reaching 10^{-2} P.A.L., and the absence of Precambrian fossils is a result of the absence of advanced organisms.

Widespread marine photosynthesis after the opening of the Cambrian period led to further increases in atmospheric oxygen

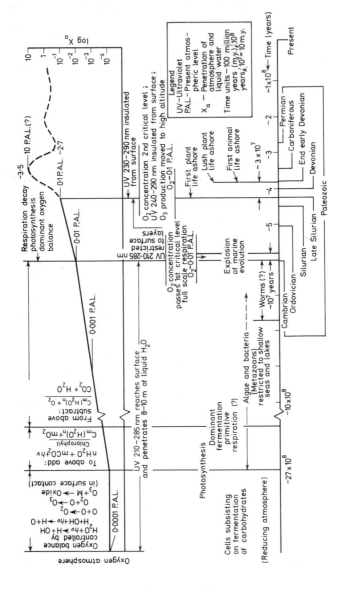

Figure 8.2. Summary chart for model of growth of oxygen in the atmosphere. (From L. V. Berkner and L. C. Marshall, Discuss. Faraday Soc. **37**, *122, Fig. 15 (1964))*

levels; with $[O_2] > 10^{-1}$ P.A.L., atmospheric ozone begins to be present in large enough concentrations to protect life *on dry land*. The geological record indicates advanced life forms ashore from the late Silurian period (420 million years ago). Evolution on dry land then caused an increased rate of photosynthesis, and the oxygen concentrations moved towards their present level. Indeed, it seems possible that the oxygen level may have overshot the P.A.L. by a factor of ten, as a result of the intense photosynthetic activity of the Carboniferous period (340 million years ago). An interesting possible consequence of this rapid photosynthesis is that atmospheric carbon dioxide would have been depleted if oxidation of organic matter could not regenerate CO_2 at a sufficient rate. The temperature of the Earth depends to no small extent on atmospheric CO_2 concentrations: carbon dioxide absorbs infra-red from the sun, and the thermal energy is trapped by the so-called 'greenhouse' effect of CO_2. A reduction in atmospheric $[CO_2]$ would therefore cause a drop in the Earth's temperature and a consequent decrease in rate of photosynthesis leading ultimately to reduction in atmospheric $[O_2]$. The cooling of the Earth was probably responsible for the Ice Ages of the Permian period (280 million years ago). Finally the CO_2 level would become restored to a higher value as a result of the decreased rate of its consumption in photosynthesis. Oscillations in CO_2 and O_2 levels may still be occurring (with a period of about 100 million years), although the effects are likely to be less pronounced than those following the first period of lush plant life ashore.

Figure 8.2 summarises the model of growth of oxygen in the atmosphere which we have presented in this section.

To a large extent, our survival depends, no less than our evolution did, on our protection by atmospheric ozone from short-wavelength solar radiation. Further, the ultimate source of energy for many chemical reactions occurring in the atmosphere is the absorption of sunlight by ozone. Much interest attaches, therefore, to the measurement of present-day concentrations and distributions of ozone in the atmosphere. Rocket and satellite experiments now make possible direct determination of the ozone altitude profile, and these determinations have come at a time when laboratory investigations allow prediction of ozone distributions on the basis of hypothetical reaction schemes. Figure 8.3, curve (a), shows the results of a typical rocket investigation of atmospheric ozone concentration. The concentration reaches a maximum at an altitude around 30 km which is quite sharply peaked (note that the abscissa

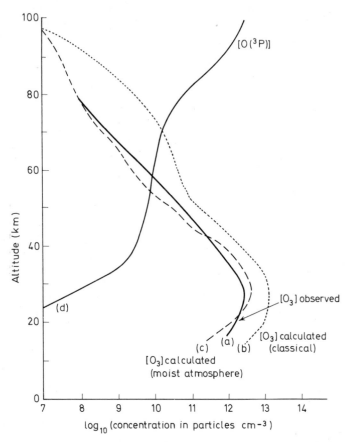

Figure 8.3. Altitude profiles for ozone and atomic oxygen concentrations: (a) experimental results (rocket and satellite) for [O₃]; (b) [O₃] values calculated from 'classical' scheme of reactions (8.2)–(8.6); (c) [O₃] values calculated for damp atmosphere; and (d) [O] values calculated for damp atmosphere. (Data of B. G. Hunt, J. Geophys. Res. **71**, 1385 (1966))

is a logarithmic scale), and atmospheric ozone is frequently described as consisting of a 'layer' centred on 30 km.

There is little doubt that the basic processes which establish the ozone layer are the reactions

$$O_2 + hv \longrightarrow O + O \text{ for } \lambda < 242 \cdot 4 \text{ nm} \tag{8.4}$$

$$O_3 + hv \longrightarrow O_2 + O \text{ for } \lambda < 1180 \text{ nm} \tag{8.5}$$

$$O + O_2 + M \longrightarrow O_3 + M \tag{8.3}$$

$$O + O_3 \longrightarrow 2O_2 \tag{8.6}$$

$$O + O + M \longrightarrow O_2 + M \tag{8.2}$$

The rate constants for reactions (8.2), (8.3) and (8.6) can be combined with calculated rates of O_3 and O_2 dissociation for an overhead sun, and the ozone concentration can be predicted at any altitude. Curve (b) of Fig. 8.3 shows that these predicted concentrations have the same general form as the observed values, and account for an ozone layer. The concentrations are, however, all higher than the true atmospheric ones, and the predicted total ozone is 0·81 cm STP as against a typical observed value of 0·25 cm STP. Several modifications of the basic scheme (8.2)–(8.6) have been suggested in order to resolve the discrepancy. The most successful of these has been the recognition that electronically excited atomic oxygen, $O(^1D)$, may react with H_2O in a moist atmosphere to yield hydroxyl radicals: the ^1D atomic oxygen is produced by the ultra-violet photolysis of ozone

$$O_3 + hv_{UV} \longrightarrow O_2 + O(^1D) \tag{8.5a}$$

$$O(^1D) + H_2O \longrightarrow 2OH \quad . \tag{8.7}$$

The details of the subsequent processes need not concern us, although we should note that they involve a chain decomposition of ozone. Curve (c) of Fig. 8.3 was obtained by use of a complex kinetic scheme which considered the reactions of $O(^1D)$, H, OH and HO_2 as reactive intermediates as well as $O(^3P)$, and it is a much better fit than curve (b) with the experimental results of curve (a). An adequate kinetic scheme for atmospheric ozone chemistry is needed in the prediction of atmospheric atomic oxygen concentrations, since reaction (8.5), together with reaction (8.4), constitute the major source of atomic oxygen. Such predictions are of particular importance because no experimental technique is yet able to provide unequivocal direct measurements of atmospheric O atom concentrations. Curve (d) of Fig. 8.3 shows the

O atom altitude distribution calculated from the 'moist atmosphere' reaction scheme used to derive curve (c).

Fears have been expressed that man might in some way upset the balance which gives rise to the ozone layer. For example, a large nuclear explosion might punch a hole in the layer and expose the earth to destructive ultra-violet radiation. It would appear, however, that the rates of the several chemical reactions are large enough to re-establish the ozone layer before much damage could be done.

8.3 POLLUTED ATMOSPHERES

A more immediate and definite problem of man's interference with his environment concerns air pollution. The chemistry of all polluted urban atmospheres necessarily involves some photochemical processes. However, in this section we wish to describe a form of pollution whose origin is essentially photochemical – the photochemical 'smog' found typically in Los Angeles.

There are several reasons why Los Angeles suffers particularly from photochemical smog. As we shall see later, the pollution derives largely from automobile exhaust gases. Los Angeles has the world's greatest traffic density, and sunshine is consistently intense, so that smog formation is favoured. In addition, the meteorological features of the Los Angeles basin, surrounded as it is by a ring of high mountains and the sea, lead to stagnation of the air and trapping of pollutants.

The characteristic pollutants are ozone and nitrogen dioxide, together with a host of organic compounds. Concentrations of O_3 and NO_2 are so high that the ozone can easily be detected by smell, and the NO_2 is apparent from the brown colour of the air. Damage to materials such as rubber, damage to vegetation, reduction in visibility, and increased incidence of respiratory disease are recognised consequences of the pollution; the most immediately obvious effect of photochemical smog is eye irritation caused by substances such as formaldehyde, acrolein and peroxyacylnitrates (PAN).

Figure 8.4 shows the diurnal variations in concentrations of several pollutants in Los Angeles, averaged for 1957. Nitric oxide is initially present, but is oxidised *after dawn* to nitrogen dioxide; ozone appears only after most of the nitric oxide has been oxidised. Small amounts of nitric oxide are known to be liberated, together with hydrocarbons, in automobile exhaust gases, and laboratory studies of the effect of ultra-violet irradiation of such gases in air show concentration–time dependences (Fig. 8.5) qualitatively similar to those for urban air pollution (Fig. 8.4). The presence of

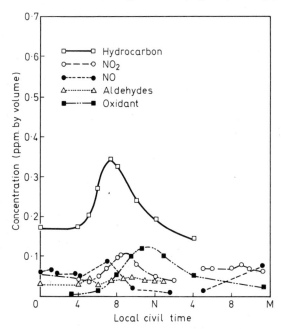

Figure 8.4. Averaged diurnal variation of pollutant concentrations (Los Angeles, 1957). (Modified from A. P. Altshuller and J. J. Bufalini, Photochem. Photobiol. **4,** 97, Fig. 13 (1965))

ultra-violet radiation is seen in both cases to be necessary to effect oxidation of NO and of hydrocarbons.

The third-order process

$$2NO + O_2 \longrightarrow 2NO_2 \qquad (8.8)$$

would be far too slow, at the low NO concentrations involved, to account for much NO_2 formation, and the only reasonable chemical reaction which can lead to sufficiently rapid oxidation of NO is

$$NO + O_3 \longrightarrow NO_2 + O_2 \qquad (8.9)$$

The absence of O_3 until almost all the NO has been oxidised

(Figs. 8.4 and 8.5) is consistent with reaction (8.9) being of major importance, at least for loss of O_3. There is, however, one major difficulty associated with the oxidation of NO to NO_2 in the purely inorganic system NO, NO_2, O_2, O_3. The only likely source of O_3 is the O atom recombination

$$O + O_2 + M \longrightarrow O_3 + M \tag{8.3}$$

At the altitudes involved in air pollution chemistry (say from ground level to at most a few km), only the photolysis of NO_2

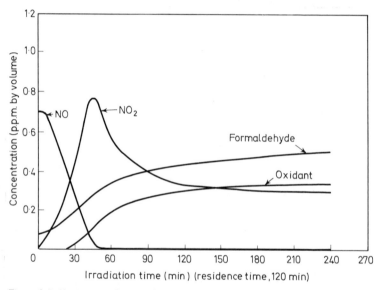

Figure 8.5. Variation with time of concentrations of reactants and products in irradiated automobile exhaust gas–nitric oxide mixtures. (From A. P. Altshuller and J. J. Bufalini, Photochem. Photobiol. **4**, 97, Fig. 11 (1965))

(λ shorter than about 400 nm) could lead to production of appreciable O atom concentrations

$$NO_2 + h\nu \longrightarrow NO + O \tag{8.10}$$

since wavelengths shorter than about 300 nm are filtered out by the ordinary atmospheric ozone layer at 30 km. Hence, we are led to the paradox that oxidation of nitric oxide requires the presence of ozone while the formation of ozone requires that nitrogen dioxide should already have been formed. The resolution of the paradox must lie in the presence of organic contaminants in the polluted atmosphere. Laboratory studies show conclusively that irradiation of nitric oxide–oxygen–olefin mixtures leads to photo-oxidation of NO. Small numbers of O atoms produced by photolysis of *traces* of NO_2 can initiate a chain oxidation which includes the steps

$$O + \text{hydrocarbons} \longrightarrow RCO + R'CH_2 \tag{8.11}$$

$$RCO + O_2 \longrightarrow RCO_3 \tag{8.12}$$

$$RCO_3 + NO \longrightarrow RCO_2 + NO_2 \tag{8.13}$$

$$RCO_2 + NO \longrightarrow RCO + NO_2 \tag{8.14}$$

$$RCO_2 + O_2 \longrightarrow RCO + O_3 \tag{8.15}$$

$$RCO_3 + NO_2 \longrightarrow RCO_3NO_2 \text{ (peroxyacylnitrate)} \tag{8.16}$$

$$R'CH_2 + O_2 \longrightarrow R'CH_2O_2 \tag{8.17}$$

$$R'CH_2O_2 + NO \longrightarrow R'CH_2O + NO_2 \tag{8.18}$$

$$R'CH_2O + NO_2 \longrightarrow R'CH_2ONO_2 \text{ (alkyl nitrate)} \tag{8.19}$$

Recently it has been recognised that electronically excited singlet oxygen (probably in the $^1\Delta_g$ state) could play an important part in the oxidation of NO to NO_2. Energy transfer from a triplet donor (e.g. excited hydrocarbon or carbonyl compound) to ground state O_2 can excite singlet O_2:

$$O_2(^3\Sigma_g^-) + D^*(T_1) \longrightarrow O_2^*(^1\Delta_g) + D(S_0) \tag{8.20}$$

(cf. Chapter 5, p. 133), and $O_2(^1\Delta_g)$ reacts with olefins to yield peroxidic compounds (cf. Chapter 6, p. 190) which could undergo thermal decomposition to radical fragments such as RCO. Thus the excitation of singlet O_2 by energy transfer can initiate the chain oxidation processes (8.12)–(8.15), and in some ways this mechansim is more satisfactory than the one relying on traces of NO_2 to initiate the chain.

Aerosols of particulate matter are found in many kinds of air pollution: most obviously, of course, in the contamination associated with burning coal (e.g. in London before the 'Clean Air Act'), but also in photochemical smog. The presence of suspended particles in the air leads to a serious reduction in visibility. Although the origin of the particulate matter in photochemical smog is not clear, it appears to involve the oxidative polymerisation of hydrocarbons (possibly aromatic); laboratory studies have shown that aerosols can be formed by the irradiation of automobile exhaust gases. Aerosols are also produced in a form of natural air pollution, found in many parts of the world, but notably in the southwestern USA. A photochemically induced contamination of the atmosphere by particulate matter gives rise to a haze or smokiness over regions possessing high densities of trees such as pines or citrus fruit. Terpenes can be oxidised by ozone to give particulate matter,† and it seems that the atmospheric aerosols are formed in this way by reactions of terpenes liberated from the trees. The ozone itself may well be produced in the non-urban lower atmosphere (i.e. where there is little or no NO and NO_2) by some process initiated by the transfer of photochemical energy from the terpene to oxygen.

8.4   PHOTOSYNTHESIS

Photosynthesis is perhaps the most important of the many interesting photochemical processes known in biology; not only was the evolution of the earth's atmosphere dependent on it, but also animal life derives energy from the sun via photosynthesis by eating plants. In this section we shall discuss photosynthesis with respect to green plants, although it should be noted that there are certain other photosynthetic organisms (e.g. some bacteria) in which the essential photochemistry may be somewhat modified.

From the point of view of organic synthesis, the over-all process consists of the formation of carbohydrates by the reduction of carbon dioxide:

$$nCO_2 + nH_2O \xrightarrow{h\nu} (CH_2O)_n + nO_2 \tag{8.21}$$

The essence of the process is the use of photochemical energy to split water and, hence, to reduce CO_2. Molecular oxygen is liberated in the reaction, although it appears at an earlier stage in the sequence of steps than the reduction of CO_2. True photochemical processes appear to produce compounds of high chemical potential which can 'drive' the synthetic sequence from CO_2 to carbohydrate in a cyclic fashion. Calvin has elucidated the mechanism of the actual carbon cycle —research for which he received the Nobel

---

†The formation of particulate matter in this way can be demonstrated dramatically in the laboratory by squeezing a piece of orange *peel* near a flask of ozonised oxygen—a bluish cloud appears in the flask.

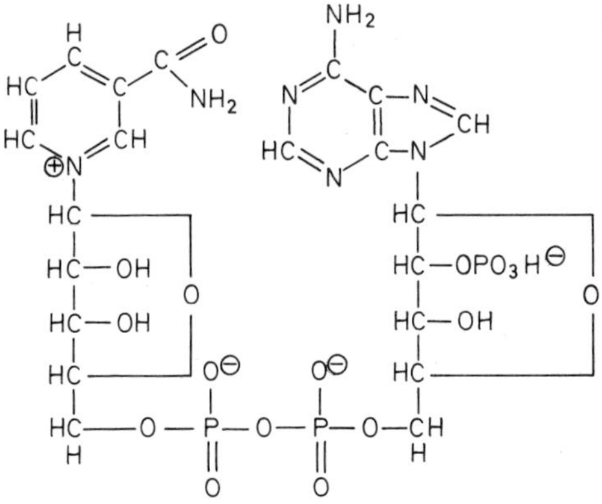

Figure 8.6. Formulae of (a) triphosphopyridine nucleotide (TPN); and (b) adenosine triphosphate (ATP)

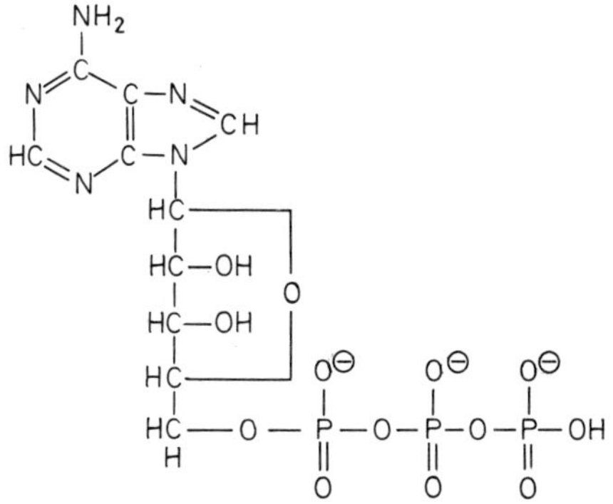

Prize. The details of the cycle do not concern us; the important feature for our purpose is the input of energy (and reducing power) by the specific compounds adenosine triphosphate (ATP) and the reduced form (TPNH) of triphosphopyridine nucleotide (TPN). Figure 8.6 shows the formulas of ATP and TPN. The synthetic carbon cycle can, in fact, be driven by ATP and TPNH in the presence of all initial enzymes and substrates, *but in the absence of light.* Thus the primary and secondary photochemical acts appear to result ultimately in the formation of ATP and TPNH by the photophosphorylation of adenosine diphosphate (ADP) and the reduction of TPN. We may represent these processes by the non-stoichiometric equations

$$ADP + \text{inorganic phosphate} \xrightarrow{h\nu} ATP \qquad (8.22)$$

$$TPN + H_2O \xrightarrow{h\nu} TPNH + O_2 \qquad (8.23)$$

It is well known that ATP is an 'energy-rich' phosphate used in many biochemical processes; the storage of chemical energy arises from the energy available in hydrolysis of ATP to ADP and H_3PO_4 (about 7–8 kcal (29–33 kJ) mol^{-1}). Since reaction (8.22) can occur independently of CO_2 reduction, and in an anaerobic environment, it seems possible that the original development of the use of light by organisms was primarily for the storage of energy rather than for the synthesis of new organic matter. The development of photosynthesis proper would then have been a later evolutionary step.

The trapping and use of solar radiation depends on the presence of chlorophyll in the plant. Figure 8.7 shows the structure of the most ubiquitous chlorophyll, chlorophyll-a. The resonance of the conjugated system brings the optical absorption into the visible region of the spectrum, at wavelengths where the solar intensity is highest at ground level. At the same time the stability conferred by the porphyrin structure ensures that absorption of radiation is followed by energy transfer or radiative processes rather than by dissociation of the chlorophyll; chlorophyll is an exceptionally efficient photosensitiser because of its ability to trap the energy of radiation and pass it on from one molecule to another until conditions are favourable for the sensitised reaction. In organic solutions the fluorescence yield is about 0·3 (although in the natural

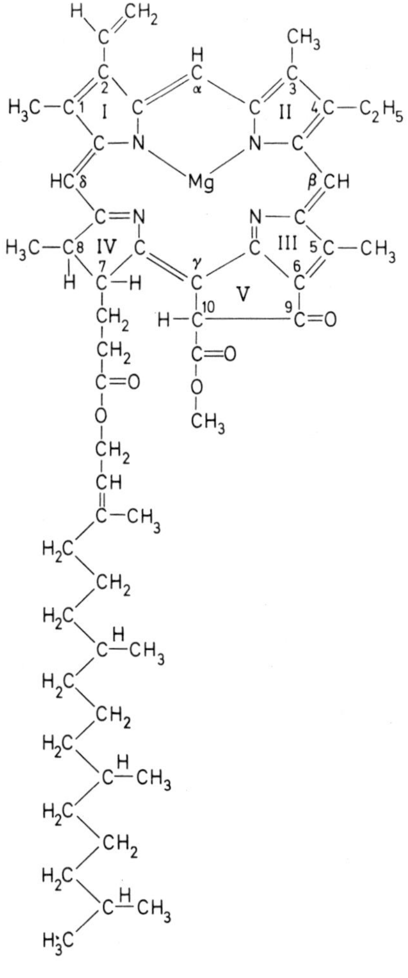

Figure 8.7. Formula of chlorophyll-a; in chlorophyll-b the 3-methyl group is replaced by —CHO

state it is much less), which is further evidence for the stability of the molecule.

The absorption spectrum of chlorophyll-a in organic solvents shows two major and two minor absorption peaks. One of the major peaks lies in the blue-violet and the other in the red region of the spectrum. In photosynthetic organisms, the chlorophyll-a is usually accompanied by one or more auxiliary pigments whose function may be to absorb radiation at wavelengths between the chlorophyll-a peaks; chlorophyll-b (chlorophyll-a with the 3-methyl group replaced by—CHO) is probably the most important absorbing auxiliary pigment in the higher plants. The auxiliary pigments appear always to transfer their excitation energy to chlorophyll-a. Red fluorescence ($\lambda_{max} \sim 680$ nm) of chlorophyll-a alone is seen in mixtures of it with auxiliary pigments, even though the latter substances absorb the incident radiation; this same red fluorescence is also observed on irradiating chlorophyll-a in its blue-violet absorption region. Thus the maximum amount of energy available in excitation of a single chlorophyll molecule is not more than 45 kcal (188 kJ) mol^{-1}. Photophosphorylation (reaction 8.22) is

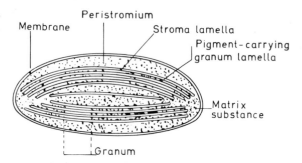

Figure 8.8. A grana-carrying chloroplast. Other photosynthetic structures are known, but this form is the most important in higher plants. (Modified from J. B. Thomas, Primary photo-processes in biology, p. 128, Fig. 40, North-Holland, Amsterdam (1965))

therefore energetically possible. However, reduction of TPN to TPNH (i.e. electron transfer from H_2O to TPN) requires about 58 kcal (244 kJ) mol^{-1}, and it is necessary to postulate some kind of 'uphill' process, possibly involving several excited chlorophyll molecules (cf. Section 5.5, p. 154). The actual energy requirement for reduction of 1 mol CO_2 to carbohydrate is near 120 kcal (502 kJ) and the quantum demand for the overall process seems to be about 6 (although values between 4 and 12 quanta per mol CO_2 have been quoted), corresponding to a total energy of about 240 kcal (1004 kJ) mol^{-1}, so that the efficiency of the multiple-quantum processes may be as high as 50%. Some rather special and efficient 'uphill' mechanism must therefore be operating in the photosynthetic plant.

Chlorophyll in the higher living plants is always to be found associated with lipoprotein membranes which are organised into a highly ordered structure known as the chloroplast. Several compounds other than chlorophyll are also found in the chloroplast. These include the auxiliary pigments, and carotenoid compounds which can both act as auxiliary pigments and protect chlorophyll against oxidative degradation (cf. p. 192). Also present are quinones (e.g. plastoquinone, α-tocopherol quinone, vitamin K) and proteins known as cytochromes. We shall see later that the quinones and cytochromes play roles in photosynthesis as important as those of the auxiliary pigments and carotenoids.

The membrane systems within the chloroplast seem to consist of a number of flattened sacs, which periodically approach each other closely to form the so-called 'grana' (see Fig. 8.8). These grana have been separated from the background 'stroma' of the chloroplasts; the photochemical behaviour of grana and stroma are different. Illumination of the grana results in liberation of O_2, but not in fixation of CO_2. It appears, therefore, that electron transfer from H_2O to TPN is associated with the membrane system. while the CO_2 reduction occurs in the stroma. There is good evidence, in fact, that the chlorophyll is attached to the lipoprotein membranes and is concentrated in the grana of chloroplasts.

Efficient photosynthesis appears to require the simultaneous excitation of more than one photosynthetically active pigment, a result which suggests the possibility of two major processes in the energy-conversion reaction of photosynthesis. The quantum efficiency of photosynthesis drops at wavelengths longer than the red absorption maximum (the Emerson or 'red-drop' effect), even though absorption in this region (675 nm–720 nm) should still populate $S_1^{v=0}$ of the chlorophyll-a. However, if supplementary light of shorter wavelength ($\lambda < 670$ nm) is added to the irradiating beam, the quantum yield for photosynthesis is considerably increased. The low quantum yields obtained with long-wave radiation can, indeed, be restored to the 'normal' values by the simultaneous illumination with short-wavelength radiation.

Figure 8.9 shows that the long-wavelength chlorophyll absorption in a chloroplast is shifted to the red, probably partly as a result of complex formation with proteins. More detailed examination of absorption spectra reveals at least two forms of chlorophyll-a in the chloroplast, which may possibly be chlorophyll-a complexed to different proteins, or present as monomers and dimers. In the absence of more specific information about the two forms of chlorophyll-a, they are known as 'pigment systems 1 and 2' (PS1 and PS2). PS2 absorbs at shorter wavelengths than PS1,

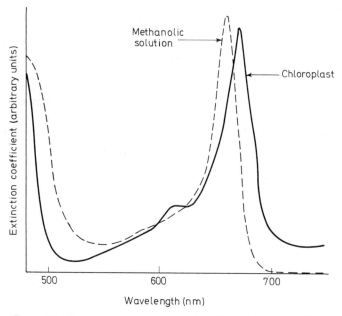

Figure 8.9. The long-wave absorption band in chlorophyll from Aspidistra eliator *in the chloroplast and in a methanolic extract. (From J. B. Thomas,* Primary photoprocesses in biology, *p. 81, Fig. 23, North-Holland, Amsterdam (1965))*

probably since absorption occurs in an auxiliary pigment (e.g. chlorophyll-b in green plants): fluorescence studies indicate, however, that the excitation always resides on the chlorophyll-a and not on the auxiliary pigment. It has proved possible to effect some separation of the two chlorophyll-containing systems by fractionation. For example, detergent treatment of chloroplasts followed by differential centrifugation produces particles of varying size containing chlorophyll-a. The lighter particles are unable to produce oxygen on irradiation in a suitable substrate, although they are more active than the heavier particles in the photo-reduction of TPN. On the other hand, the heavier particles are able to liberate oxygen; these particles are found to be enriched in chlorophyll-b. The clear indication is that the lighter particles consist exclusively of PS1, and the heavier ones of both PS1 and PS2.

One scheme suggested for the formation of ATP and TPNH is illustrated in Fig. 8.10.† In outline the mechanism consists of two partial oxidation and reduction reactions. Absorption of light by PS2 leads to the formation of a strong oxidant (eventually yielding molecular oxygen) and a weak reductant. The photoact driven by PS1 generates a strong reductant and, concomitantly, a weak oxidant. The weak reductant and weak oxidant of PS2 and PS1 reactions, respectively, are linked by a chemical bridge of compounds involving quinones, cytochromes and a copper-containing protein, plastocyanin. TPNH is derived from the strong reductant finally produced, and ATP is generated in the segment of the chain joining the two photoacts.

Known redox potentials indicate possible identities for the several oxidants and reductants; some reasonable compounds are indicated in Fig. 8.10. The initial process appears to involve oxidation of hydroxyl ions from water according to

$$2OH^- \longrightarrow \tfrac{1}{2}O_2 + H_2O + 2\varepsilon \qquad (8.24)$$

and the first electron acceptor is plastoquinone, which acts, via the 'bridge', as a donor to the PS2 redox system; the starting compound of this latter system is probably cytochrome-f. Recent evidence suggests that the strong reductant produced may be the reduced form of ferredoxin, an iron-containing protein found in the cytoplasm which has the relatively high negative reduction potential of -0.43 V in this system.

The function of the chlorophyll in photosynthesis is now becoming clear: under the influence of light it can cause electron transfer and bring about oxidation–reduction changes. We may finally consider the experimental evidence which confirms the ability of chlorophyll to behave in this way.

†The scheme is not universally accepted (see, for example, ARNON, D. I., *Experientia*, **22**, 1 (1966)). It does not, in any case, describe the precise mechanism of the phosphorylation and reduction, but rather the energetics of the processes.

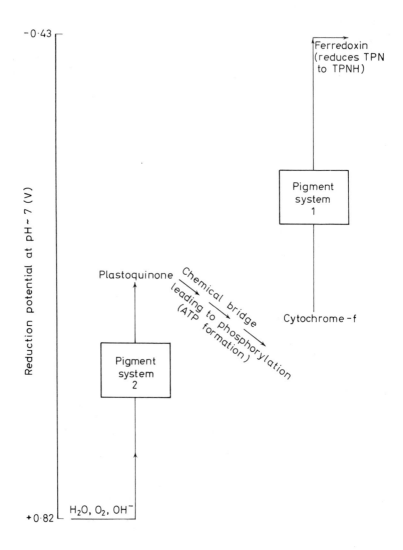

*Figure 8.10. A suggested scheme for the photochemical formation of ATP and TPNH illustrating the approximate reduction potentials of the various components. (Modified from a more complex scheme given by L. R. Vernon and M. Avron, Ann. Rev. Biochem. **34**, 269, Fig. 1 (1965))*

Electron spin resonance spectra have been obtained from illuminated chloroplasts. Although the detailed interpretation of the ESR responses is uncertain, it is clear that free radicals or electrons are produced on irradiation of chlorophyll. Since chlorophyll itself is not decomposed, the inference is that chlorophyll provides a source of electrons on photoexcitation. The optical absorption spectra of chloroplasts show that chlorophyll is not crystalline *in vivo*, but electric dichroism measurements indicate that there is some orientation of chlorophyll and it is possible that the concepts of solid state physics are applicable to the photosynthetic apparatus. This means, in particular, that the excitation of chlorophyll is to be looked on as the generation of an electron–hole pair, and that the layered structures of the chloroplasts are important because they allow electrons and holes to become separated from each other by electron transfer. Migration of electrons can occur over long ranges (cf. Section 5.4), and the excitation can therefore persist until it reaches an electron-acceptor site (e.g. plastoquinone or ferredoxin). The positive holes in the chlorophyll are filled from suitable electron donors (hydroxyl ions or cytochrome-f). Photoconductivity has been observed in chloroplast layers, and provides further evidence that electron migration occurs in illuminated chloroplasts.

This section has shown that the mechanism of photosynthesis is understood in outline. Elucidation of the details of the complex photophysical primary process may well further our general understanding of photochemistry.

8.5   VISION

Three phyla (arthropods, molluscs and vertebrates) have developed well-formed eyes, although the anatomy and evolutionary development of vision in the three phyla is entirely different. It is therefore remarkable that the photochemistry of the visual process is nearly identical for the three types of eye. Vision is stimulated in each case by the photochemical transformation of a pigment containing a moiety related to Vitamin A (retinol).† It is this photochemistry with which we are mainly concerned, although the photoreceptive

---

†There are two vitamin A compounds, A_1 and A_2; A_2 contains a second double bond at the 3:4 position in the ring (see Fig. 8.11). In this section 'vitamin A' refers to A_1; A_2 may be important in the visual processes of some fish.

structures of the eye must be discussed in so far as they affect the photochemistry. Colour vision is not treated; however, the basic photoprocess is that to be described, and colour perception may depend in addition on the presence of three pigments (chlorolabe, erythrolabe and cyanolabe) sensitive to different spectral regions.

The gross anatomy of the vertebrate eye — in particular, the

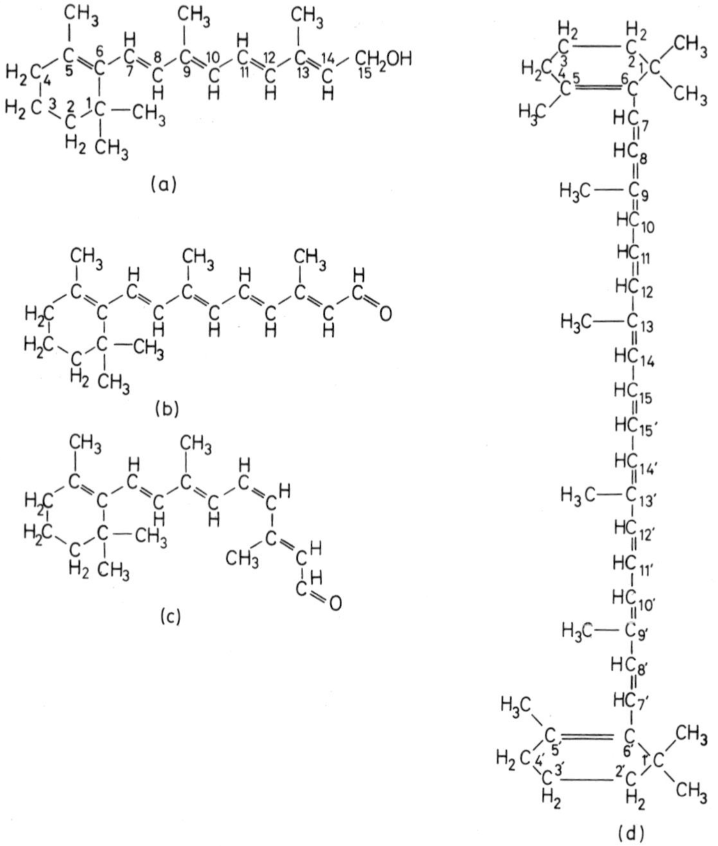

Figure 8.11. Formulae of (a) retinol, (b) all-trans-retinal (c) 11-cis-retinal (d) β-carotene

system of lens and retina — is too well known to need description here. The receptors of the retina consist of 'rods' and 'cones': the former possess high sensitivity, and are used at low light intensities, while the latter are less sensitive but may carry colour-selectivity. Electron microscopy has revealed the structure of the rods and cones for some species, and Fig. 8.12 is a diagram of the rod and cone outer segments of the *Necturus* eye. A number of lamellae are formed by the infolding of plasma membranes, and these lamellae are the carriers of the visual pigments. Observations of the rod outer segments of the frog have shown that illumination transforms the straight cylinders into crumpled structures with many transverse 'breaks', as if the lamellar discs had fallen apart. The effect is consistent with a light-induced structural change in the visual pigment.

There is evidence that a single quantum of radiation can stimulate a retinal rod. The absorption of one quantum does not, however,

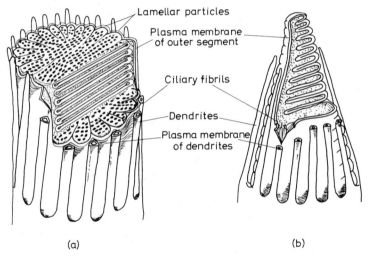

Lamellar particles

Plasma membrane of outer segment

Ciliary fibrils

Dendrites

Plasma membrane of dendrites

(a)   (b)

Figure 8.12. Sketch of (a) rod and (b) cone outer segments of the eye of Necturus. *(From C. Wald, P. K. Brown and I. R. Gibbons,* J. opt. Soc. Am. **53**, *20 (1963))*

result in vision, and several quanta (values between two and six are considered reasonable) must reach the same rod within a relatively short period. Even so, the process is remarkably efficient, and the energy of the ultimate reaction greatly exceeds that absorbed by the retinal pigment. The absorption of light appears to initiate a chain reaction which derives its energy from metabolism, and visual excitation is a result of 'amplification' of the light signal received at the retina. One possible amplification mechanism that has been suggested is that the pigment should be considered a 'pro-enzyme' whose enzymic activity is released following the light-induced reorganisation. A single molecule of enzyme can catalyse reactions of many substrate molecules, and the process described will act as a biochemical light amplifier. It should be stressed that this mechanism for the excitation process is not universally accepted, and several other possibilities have been raised. Nevertheless, as we shall see shortly, direct experimental evidence shows that retinal pigments show enhanced ATP-ase activity upon irradiation.

In 1876 Boll discovered that the rose colour of the frog's retina faded in bright light. This bleaching of the so-called 'visual purple' demonstrated explicitly the occurrence of a photochemical reaction in vision. Subsequent studies showed that the bleaching is reversible if the retina is kept *in situ*: the reversibility was lost in solutions of visual purple extracted from the retina, even though the initial photobleaching still occurred. It is now recognised that the bleaching is too slow to be responsible for the sensory visual response, and that it is the end result of a sequence of reactions involved in nerve excitation. We now turn to an examination of the nature of the pigment and its photochemistry.

Chemical studies of extracts from the retina show the visual pigments to be compounds of a carotenoid substance with a protein. Rhodopsin (visual purple), which is typical of the pigments, contains 11-*cis*-retinal as the carotenoid chromophore and the protein scotopsin. Figure 8.11 shows the relationship of retinal to retinol (Vitamin A) and to β-carotene. Animals derive their retinol from carotenoids of plant origin, and retinal is produced in the retina by enzymic oxidation of retinol. Scotopsin is the class of pigment protein found exclusively in rods (*photopsin* is found in cones); the details of the structure of opsins are not known, although it is clear that scotopsin is a protein with a molecular weight of about 40 000. The bond between retinal and opsin either results from condensation of the aldehyde with a protein amino group to form a Schiff base

$$C_{19}H_{27}HC{=}O + RNH_2 \longrightarrow C_{19}H_{27}HC{=}NR + H_2O \qquad (8.25)$$

or it may involve the reaction of the —CHO group with a protein —SH group to form a sulphur bridge. The photo-physiological properties of rhodopsin depend on the complexing between chromophore and protein. First, the optical absorption in the red ($\lambda_{max} \sim 500$ nm) apparently derives from a red-shifted transition in the 11-*cis*-retinal (λ_{max} in ethanol ~ 378 nm). Secondly, the colour changes (these include bleaching; see below) observed on illumination of rhodopsin do not occur with retinal on its own. Thirdly, carotenoids are unlikely to initiate nervous stimulatory responses, and it is hard to see how they could take part in an amplification process.

Irradiation of rhodopsin leads to the formation of a series of derivatives of varying stability and colour; the processes may be described by the sequence

$$
\begin{array}{ll}
\text{rhodopsin} \xrightarrow{h\nu} \text{prelumirhodopsin} \xrightarrow{\text{thermal}} \text{lumirhodopsin} \\
\quad \text{(red)} \qquad\qquad\quad \text{(red)} \qquad\qquad\qquad\quad \text{(orange-red)} \\
\\
\qquad\qquad\qquad\qquad\qquad\qquad\qquad\qquad\qquad\quad \Big\downarrow \text{thermal} \\
\text{retinal} + \text{opsin} \xleftarrow{\text{thermal}} \text{metarhodopsin II} \xleftarrow{\text{thermal}} \text{metarhodopsin I} \\
\quad \text{(colourless)} \qquad\qquad \text{(yellow)} \qquad\qquad\qquad\quad \text{(orange)}
\end{array}
$$

$$(8.26)$$

The primary photochemical process appears to be photoisomerisation of 11-*cis*-retinal to all-*trans*-retinal. Current opinion ascribes the subsequent chemical changes to the inability of the straight all-*trans*-compound to be sterically accommodated on the surface of the opsin, only the bent 11-*cis*-retinal 'fitting into' the protein, and photoisomerisation leads first to strained structures and ultimately to cleavage of the protein—chromophore bonds. The changes in optical absorption are consistent with this view. Lowering of the energy of the excited state due to interaction of retinal with opsin leads to a red shift, and the stronger the interaction the larger the shift. Thus, as the progressively more strained structures lumirhodopsin and the metarhodopsins are formed, the shift becomes smaller and the absorption maximum moves towards the blue.†

The electronic states involved in the *cis–trans* isomerisation have not been established. Although a transient spectrum consistent with a triplet species is observed on the flash photolysis of all-*trans*-retinal, there is no evidence for triplet states in the isomerisation in rhodopsin. It is possible that the $^3(n, \pi^*)$ levels lie above the first excited singlet state (possibly a π, π^* state) in the protonated Schiff base, but below it in the free aldehyde, so that triplets are formed only in free retinal.

---

†The blue shift is not observed on formation of prelumirhodopsin, and there may actually be a red shift; it has been suggested that the ground state of prelumirhodopsin lies at a higher energy than that of rhodopsin as a result of the strained geometry, while the excited states of both compounds lie at similar energies.

As we have noted earlier, the final dissociation of irradiated rhodopsin into free retinal and protein is too slow to account for visual excitation. However, two new —SH groups are exposed at the metarhodopsin I stage, and a proton binding group with a pK of about 6·6 at the metarhodopsin II stage, and it may be this form of the protein (with the chromophore still attached) that is responsible for excitation. Experiments on the enzymic activity of rhodopsin have shown that it can catalyse phosphorylation (i.e. it is an ATP-ase), and the activity is greatly enhanced on exposure to light. Further, the wavelength for greatest enhancement coincides with that of maximum optical absorption by rhodopsin. It is interesting that light can cause all-*trans*- → 11-*cis*-retinal isomerisation, as well as the reverse reaction, and mixtures of all-*trans*-retinal with opsin show ATP-ase activity only after irradiation: the activity is proportional to the amount of rhodopsin formed. Thus there is much circumstantial evidence to suggest that the visual process involves the release of enzymic activity on irradiation of rhodopsin. What is not yet known is how this activity is transferred to electrical nerve impulses.

8.6 PHOTOGRAPHY

The making of a more or less permanent record of areas of light and shade by photography represents the best-known of all 'applied' photochemical processes. By far the most common form of the photographic process (both for monochrome and for colour photography) is based on silver halides as the photosensitive materials. Microcrystalline grains of silver halide suspended in gelatin are coated onto a suitable support (film, glass plate, paper, etc.) to form the light-sensitive 'emulsion'. Prolonged exposure to light causes darkening of the emulsion — the *print-out effect* — which X-ray powder patterns show clearly to be a result of metallic silver formation. Much shorter exposures produce a so-called *latent image* in the silver halide grains. This latent image may be turned into a visible silver deposit by a 'developer', which is a suitable reducing agent. All developers are, in fact, thermodynamically reducing towards silver halides, and the presence of the latent image seems to lead to an increased rate of reduction to metallic silver rather than to a change in the ultimate reducibility of the emulsion. Extended development of unexposed emulsion leads finally to darkening ('fog'), so that the discrimination by development between exposed and unexposed areas depends on the difference in reduction rate for the two areas.†

Several experiments give direct evidence that the latent image consists of metallic silver present in the halide grains, but at much lower concentrations than in the print-out image. Using techniques capable of detecting optical density changes around 10^{-6}, it is possible to find measurable silver densities in latent images even at the threshold of developable exposures. There is also a marked similarity in the influence of environmental factors (e.g. the presence of electric fields, or of crystal imperfections — see below) on the location of print-out silver particles and of development centres. Our discussion of the primary photochemical processes will therefore deal mainly with the production of print-out silver, and it will be assumed that latent image formation is photochemically identical but that it involves much lower conversions. However, one important feature of latent image formation is the decrease in emulsion sensitivity at very low light intensities ('Reciprocity-law failure') which indicates the presence of a multi-quantum process. There is evidence that a single silver atom is generally unstable in the halide lattice, having a lifetime of only a few seconds, and that at least two atoms are needed unless there is some pre-existing stabilising centre.

The presence of metallic silver atoms in the latent image appears to lower the activation energy for the reduction reaction in development, and thus enhance the rate. Development, once it has been initiated at one site on a grain, proceeds with an increasing rate as more and more silver is formed, until the entire grain is developed. This autocatalytic activity of silver has been clearly demonstrated: a low concentration (about 10^{15} atoms cm^{-2}) of metallic silver evaporated onto a silver halide surface renders the surface 'developable'.

Silver halides show the phenomenon of photoconductivity, and it is believed that irradiation of silver halide raises photoelectrons from the valence band to the conduction band of the halide. The mechanism for the production of free silver then involves the migration of the photoelectrons and of interstitial silver ions to preferential sites in the halide; free silver atoms are formed by the combination of silver ions and electrons. The free silver so formed

†In normal practice the developed image corresponding to exposed and unexposed areas is rendered permanent by 'fixing': the unexposed (and, hence, unaffected) grains of silver halide are dissolved in sodium thiosulphate solution immediately after development.

acts as an efficient trap for photoelectrons produced subsequently, and so further silver ions are discharged *near the same place* as the initial atom. Specks of silver grow, therefore, at the original preferential site. The positive 'holes' left behind by the electrons may have some mobility, so that they can diffuse towards the surface of the halide grain and liberate free halogen.

Two forms of the general mechanism have been proposed, which differ in the sequence of events. Figure 8.13 represents a mechanism essentially the same as that given by Gurney and Mott; Mitchell's alternative scheme involves initial trapping of an electron by an Ag^+ ion, and subsequent adsorption of Ag^+ on the growing silver speck to trap further electrons. In either case the basic processes are similar. The steps up to the formation of the two-atom speck are reversible, which is consistent with the experimental evidence for latent image stability only with aggregates of more than two atoms (see above).

The hypothetical Gurney–Mott (or Mitchell) mechanism has been well-substantiated by experiment. The photoconductivity of silver halides *previously darkened by irradiation* is less than that of undarkened halides, which indicates that the colloidal silver particles (or the physical imperfections introduced into the lattice by their formation) are effective in trapping electrons. The participation of charged species in image formation has been shown in an experiment in which a crystal of silver chloride is held between two electrodes, and exposed to radiation through a semi-transparent conducting aperture in one electrode. Strongly absorbed light is used, and in the absence of an electric field, silver image formation is restricted to a region near the crystal surface. If, however, a strong electric field is applied,† with the illuminated electrode made negative, then photoelectrons are displaced towards the interior of the crystal and the region of darkening is similarly displaced, which indicates that photolytic silver separates at sites where the photoelectrons are trapped. The same effects have subsequently been shown in the microscopic grains of a silver bromide emulsion using *weakly* absorbed light. Electron microscopic examination shows that, in the absence of an applied field, silver grains are liberated uniformly throughout the grain. In the presence of the field, silver particles concentrate almost entirely near the positive side of the grain. Further, a high concentration

---

†The actual experiment involved flashes of light and a pulsed electric field to prevent significant electrolysis of silver chloride.

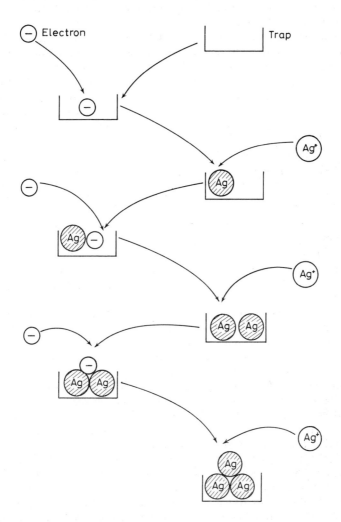

Figure 8.13. Representation of silver image formation based on the Gurney–Mott concentration principle. In Mitchell's alternative scheme, the first act is trapping of an electron by Ag^+ : the Ag formed adsorbs a second Ag^+ ion, and so on

of atomic bromine builds up near the negative side, thus demonstrating the drift of 'holes' in the grain and confirming the photoliberation of halogen.

The escape of halogen is necessary for photochemical change; unless holes diffuse to a grain surface and liberate halogen, they recombine with electrons and no free silver is formed. Investigation of the silver distribution in irradiated crystals of silver halide has shown that the silver is concentrated in a sub-surface layer which is at most a few microns thick. Further, the quantum yield for photodecomposition in large crystals of silver bromide is small when the exposing light is weakly absorbed (e.g. $\phi \sim 0.02$ at $\lambda = 436$ nm), but increases to near unity at very short wavelengths at which all the light is absorbed near the surface. It has been shown that with weakly absorbed light all liberated bromine comes from a surface layer about $\frac{1}{3}\mu$m thick, and that the quantum yield is high in that surface region.

A real crystal or grain of silver halide may possess chemical impurities and physical imperfections; in the case of emulsions, the gelatin may also enter into the photochemical processes.

The role of chemical impurities is shown clearly by experiments on silver halides with substitutional cuprous ions. The quantum yield for silver production is near unity even in the centre of a crystal, and the number of silver atoms formed at saturation level is equal to the number of cuprous ions initially present. EPR spectra show the formation of cupric ions concomitant with silver atom production, and the impurity cuprous ions appear to act, therefore, as traps for the holes in the bulk of the crystal. Flash photolysis studies confirm this interpretation. A photo-darkening is observed in both pure and cuprous-doped silver halide crystals, and the rate of build-up is similar. However, the darkening is transient in the pure crystals, and fades in a few milliseconds, whereas it is stable in the doped silver halide.

Extended crystal imperfections serve two functions in photographic image formation. First, they provide charged centres which act as traps for electrons and holes, and, secondly, they provide sites from which interstitial silver ions are readily generated. It is probable that in the volume of a halide grain the principal sites for silver separation are on internal imperfections such as jogs on edge dislocation lines, grain boundaries, and twin planes (although at the surface — which can be considered as an imperfection itself — there is no shortage of electron traps, and additional imperfections are probably unnecessary). The lifetime of photoelectrons is increased from about 1 μs to 10 μs by annealing out physical imperfections. Again, introduction of slip planes by mechanical deformation leads to preferential darkening of the deformed regions. Microscopic examination of print-out silver shows that dislocations and mosaic boundaries are effectively 'decorated' by the silver; essentially similar decoration of imperfections is seen after development of a latent image.

Gelatins containing labile sulphur or reducing groups have long been known to increase the sensitivity of a photographic emulsion, and in present-day manufacturing techniques controlled additions of sensitisers are made to inert gelatins. The exact mode of action of chemical sensitisers is not yet established, although it seems certain that silver sulphide is formed in sulphur-containing emulsions. The sulphide can then act at the sites of image centres, either to provide deeper electron traps or to increase stability during the earliest stages of image formation. Silver sulphide can also reduce recombination of electrons and holes and remove bromine, since it can capture holes or bromine.

Spectral sensitisation of silver halide emulsions can be achieved by adsorption of suitable dyes onto the halide grains. Such sensitisation is important, since it permits image formation by radiation of wavelength longer than that effective with unsensitised emulsions (say 490 nm — blue-green — for silver iodide emulsion), and it offers an excellent example of a reaction photosensitised by energy or electron transfer. Indeed, spectral sensitisation of photographic emulsions seems to have been the first recognised case of photosensitisation (1873).

Perhaps the most important class of sensitising dye is that of the cyanines: these dyes contain heterocyclic or benzenoid nuclei joined by a $=CH(-CH=CH)_n$ bridge, the π electrons of which take part in the spectral transitions leading to sensitisation. It is a characteristic of these dyes to be strongly adsorbed to the silver halide grains. The fluorescence yield of the adsorbed dyes is much lower than that of ordinary solutions. However, the phosphorescence yield is also small, and the decrease in fluorescence does not appear to result from a rate of $S_1 \rightsquigarrow T_1$ ISC enhanced by the heavy-atom effect (cf. p. 115). Rather, the results suggest that the fluorescence is quenched by transfer of excitation from the S_1 level of the dye to the silver halide. The close correspondence between dye absorption spectra and the spectral sensitivities of sensitised emulsions indicates clearly that the excitation transfer is also responsible for sensitisation of the photographic process.

The effect of the dye seems to be to make available photo-electrons in the silver halide conduction band. Photoconductivity of thin dyed crystals of silver halide is observed at wavelengths absorbed by the dye and longer than those to which the undyed crystal responds. Two main sensitisation mechanisms which fit the experimental facts can be envisaged: these are transfer from the dye to the halide either of an electron or of excitation energy. The two processes are shown schematically in Fig. 8.14(a) and (b):

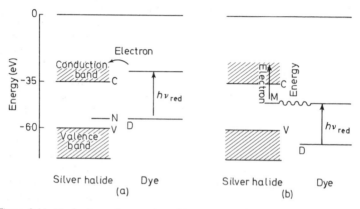

Figure 8.14. Mechanisms of spectral sensitisation: (a) electron transfer, (b) energy transfer

the energies are referred to a zero for a free electron at rest *in vacuo*. The long-wavelength limit of the unsensitised process (ca. 490 nm, equivalent to 2·5 eV or 59 kcal (247 kJ) mol^{-1}) probably corresponds to the minimum energy gap between valence and conduction bands. However, dye-sensitised image formation can be achieved at wavelengths as great as 1 300 nm (24 kcal (101 kJ) mol^{-1}), at which the energy is insufficient to excite an electron from valence to conduction bands in the halide. In both mechanisms it is supposed that energy-rich surface sites play an important part. A restriction is placed on the dye energy levels in the electron transfer mechanism (Fig. 8.14a), since the excited dye level must lie within the halide conduction band. It has been shown that this energy requirement can be satisfied for many sensitising dyes, although this does not, of course, establish the occurrence of the electron transfer process. No energy restrictions are imposed on the dye by the energy transfer process (Fig. 8.14b). In this case, however, it is necessary that there be an energy-rich surface state, M, with which the dye molecule is coupled, so that transfer of energy, equivalent to the excitation energy of the dye, can excite an electron into the halide conduction band.

Although regeneration of the dye molecule is not necessary to the sensitisation of latent image formation, it is, in fact, observed for print-out image production, and any mechanism proposed for sensitisation must account for dye regeneration. No special process is required in the energy transfer mechanism, since energy loss from the dye will restore the molecule to the ground state. On the other hand, if the electron transfer mechanism is operative, the hole in the dye molecule must be replaced, most probably by electron transfer from a halide ion of the silver halide. Transfer from a regular lattice halide ion would require an activation energy (*at least* the energy between V and D in Fig. 8.14a), such as to make the process improbable at room temperature, and it is likely that regeneration involves the passage of an electron from an energy-rich, surface, bromide ion represented by the level N in Fig. 8.14(a).

The temperature-dependence of the excitation transfer process can be measured by exposing an emulsion at reduced temperature and subsequently developing under normal conditions. There is a drop in sensitivity as the exposed temperature is reduced which must be ascribed to a reduction in efficiency of excitation transfer. The variation of sensitivity (in the long-wavelength sensitised region) with temperature does not fit the Arrhenius equation. One interpretation of the temperature-dependence has been in terms of 'tunnelling' through the potential barrier to the transfer process. Electrons can tunnel through barriers of considerable height, as a result of their small mass, and calculations of rate for electron tunnelling give a good fit with the experimental sensitivity data.

8.7 PHOTOCHROMISM

Silver halide photography involves the production of an essentially permanent optically visible effect by means of an irreversible photochemical process. The production of a *reversible* photo-induced colour change is referred to as *photochromism*. In photochromic systems irradiation alters drastically the absorption spectrum; but when the irradiation source is removed, the system reverts to its original state. The visible effect often involves the appearance of colour in a previously colourless material, although changes of colour – for example from red to green – are also known.

Numerous applications of photochromic substances have been suggested, and some of these have already entered commercial practice. Various forms of data storage are possible (including image storage – i.e. photography). Plastics incorporating a photochromic dye have been used for sunglasses, and for aircraft windows, which darken in bright sunlight but become lighter again under less intense illumination. A similar application is for aircraft canopies designed to protect pilots against the brilliant light flash from a nuclear explosion; in this case the photochromic system must develop a high optical density rapidly on illumination, but relax back to the colourless form within a few tens of milliseconds after the flash. A rather more frivolous application is in the manufacture of toy dolls which can be 'suntanned': a photochromic dye is used which produces a brown coloration on exposure to sunlight.

The major difficulty in the use of products incorporating photochromic materials is the rapid 'fatigue' exhibited by most known photochromic substances. A very small lack of reversibility soon leads to chemical decomposition, especially in applications where the illumination is sunlight. Most of the photochromic systems reported are really only able to undergo reversal a limited number of times.

In principle, photochromism could result from the photochemical production of a significant population of excited molecules, from which absorption to a higher level could occur. Although such processes are, of course, well known (e.g. for triplet–triplet absorptions), with moderate illumination intensities the steady state concentrations of excited species are too low to be useful in photochromic applications. The other main mechanisms responsible for photochromic behaviour are isomerisation, dissociation and charge-transfer (redox). Many hundreds of specific photochromic substances are known, and we must select just one example to illustrate each type of mechanism.

Many aromatic nitro-compounds exhibit photochromic isomerisation; the process is believed to involve photoisomerisation from the colourless nitro-form to the coloured aci-form:

nitro-form   aci-form   (8.27)

$R_1 = H$, CH_3, C_6H_5 etc; R_2 = electron-withdrawing group. (The aci-form must also undergo some dissociation since it is a strong acid.)

Irradiation of chromium hexacarbonyl in a plastic matrix (ca. 0.1% $Cr(CO)_6$) leads to the formation of a deep yellow colour as a result of photodissociation of the hexacarbonyl. In the plastic, CO cannot escape, and recombination occurs in about 4 h at room temperature:

$$Cr(CO)_6 \underset{\text{dark}}{\overset{\text{light}}{\rightleftharpoons}} Cr(CO)_5 + CO \qquad (8.28)$$

Both organic and inorganic charge-transfer or redox photo-chromic systems are known. A typical reversible photochemical redox reaction occurs in a mixture of mercurous iodide and silver iodide:

$$Hg_2I_2 + 2AgI \underset{\text{dark}}{\overset{\text{light}}{\rightleftharpoons}} 2HgI_2 + 2Ag \qquad (8.29)$$
$$\text{green} \quad \text{yellow} \quad \text{red} \quad \text{black}$$

8.8 OPTICAL BRIGHTENERS

Coloured fluorescent dyes and pigments find many non-scientific applications: they are used for the brilliant 'dayglo' paints, for textile dyes, and to obtain special theatrical effects. No application is so widespread, however, as the use of special fluorescent sub-stances as *optical brighteners* or *bleaches* in the 'whiter-than-white' washing powders. The principle behind the operation of an optical bleach is that the substance should absorb in the ultra-violet and radiate in the visible region, so that the washed (white) textile apparently reflects more light than was incident upon it.

A substance which is to be suitable as an optical bleach must satisfy several stringent requirements. First, it must not absorb at all in the visible, since this would lead to coloration of the fabric, but must absorb strongly in the near ultra-violet, where there is still some intensity available from natural or artificial light-sources. Secondly, the fluorescence must lie in the short-wavelength end of the visible spectrum, as otherwise the fluorescence would give an apparent undesirable yellowing to white fabric. Thirdly, the fluorescent substance must be photochemically stable, and it must not sensitise degradation or oxidation of the fibre material. Finally, the substance must be soluble or dispersible in the aqueous detergent solution, but must be sufficiently strongly adsorbed by the textile fibres for an appreciable concentration to build up during washing and to remain after rinsing.

Over 200 chemically different optical brighteners are commercially available, and the choice for a given detergent depends on the type of textile and washing conditions for which the detergent is intended. Cellulosic fibres (e.g. cottons) possess adsorption characteristics different from those of synthetic fibres, and the brightener must be selected accordingly; in many cases several brighteners may be added to the detergent to give a wide spectrum of activity.

Cellulosic fibres are hydrophilic and swell in water so that the pores in the amorphous region grow to around 15–30 Å in diameter, large enough to admit the brightener molecules. High affinity for the fibre is obtained if the brightener possesses several conjugated double bonds and aromatic nuclei of coplanar configuration: fortunately, this is also just the requirement for a high fluorescence yield (cf. Section 4.3, pp. 95–96). Almost all brighteners now used for cellulosic fibres are derivatives of bis-triazinyl-4, 4'-diaminostilbene-2,2' disulphonic acid (Fig. 8.15a): the *trans-*

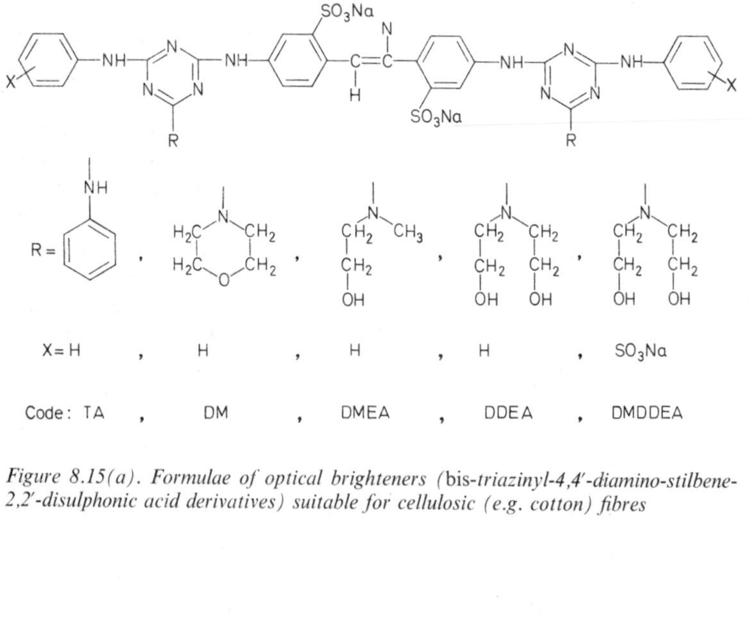

Figure 8.15(a). Formulae of optical brighteners (bis-triazinyl-4,4'-diamino-stilbene-2,2'-disulphonic acid derivatives) suitable for cellulosic (e.g. cotton) fibres

isomer is the one adsorbed. The various substituents have little effect on the emission spectra, but alter the characteristics of adsorption onto the fibre.

Brightening of hydrophobic fibres (e.g. nylon, polyester and acetate) takes place in a manner similar to the dyeing of these fibres, possibly involving the penetration of the molecules into canals between the fibre molecules, or, alternatively, as a result of actual solution of brightener in the solid fibre. Figure 8.15(b) shows some brighteners active for both polyamide and polyester fibres.

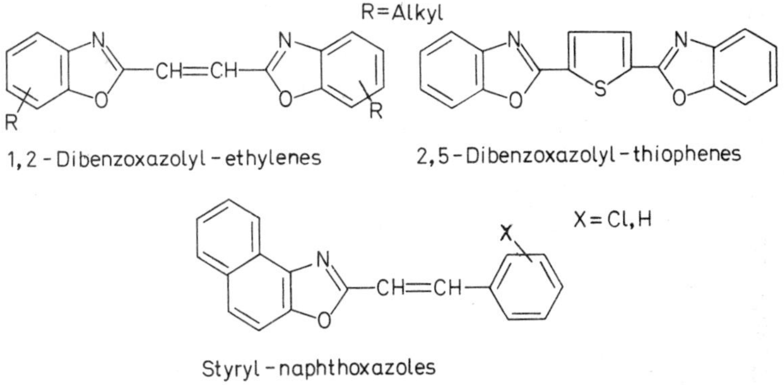

1,2 - Dibenzoxazolyl - ethylenes

2,5 - Dibenzoxazolyl - thiophenes

Styryl - naphthoxazoles

Figure 8.15(b). Brighteners suitable for polyamide (e.g. nylon) and polyester (e.g. Terylene) fibres

BIBLIOGRAPHY

L. V. BERKNER and L. C. MARSHALL, 'The history of oxygenic concentrations in the earth's atmosphere', *Discuss. Faraday Soc.* **37**, 122 (1964)

B. G. HUNT, 'Photochemistry of ozone in a moist atmosphere', *J. geophys. Res.* **71**, 1385 (1966)

P. A. LEIGHTON, *Photochemistry of air pollution,* Academic Press, New York (1961)

A. P. ALTSHULLER and J. J. BUFALINI, 'Photochemical aspects of air pollution: a review', *Photochem. Photobiol.* **4**, 97 (1965)

G. WALD, 'Life and light', *Scient. Am.* (October, 1959)

J. B. THOMAS, *Primary photoprocesses in biology:* Chapter 4, 'Photosynthesis'. North-Holland, Amsterdam (1965)

M. CALVIN, 'The path of carbon in photosynthesis', *Angew. Chem.* **2**, 65 (1962)

J. B. THOMAS, *Primary photoprocesses in biology:* Chapter 5, 'Vision'. North-Holland, Amsterdam (1965)

T. H. JAMES (Ed.), *The theory of the photographic process,* 3rd edn, Macmillan, New York (1966)

R. DESSAUER and J. P. PARIS, 'Photochromism', *Adv. Photochem.* **1**, 275 (1963)

P. S. STENSBY, 'Optical brighteners as detergent additives', *J. Am. Oil Chem. Soc.* **45**, 497 (1968)

APPENDIX 1. FUNDAMENTAL PHYSICAL CONSTANTS OF IMPORTANCE IN PHOTOCHEMISTRY

Symbol	Constant	Value
c	Speed of light *in vacuo*	2.998×10^8 m s^{-1}
h	Planck's constant	6.63×10^{-34} J s
k	Boltzmann constant	1.38×10^{-23} J K^{-1}
N	Avogadro's number	6.023×10^{23} mol^{-1}
R	Gas constant	8.31 J mol^{-1} K^{-1}
m	Mass of electron at rest	9.11×10^{-31} kg
e	Charge of electron	1.60×10^{-19} C
		$(4.80 \times 10^{-10}$ e.s.u.$)^a$

^{a}See footnote on p. 17.

APPENDIX 2. CONVERSION TABLE FOR ENERGY UNITS

A	B				
	erg molecule^{-1}	J mol^{-1}	cal mol^{-1}	eV	cm^{-1}
erg molecule^{-1}	1	6.021×10^{16}	1.439×10^{16}	6.242×10^{11}	5.035×10^{15}
J mol^{-1}	1.660×10^{-18}	1	2.370×10^{-1}	1.036×10^{-5}	8.361×10^{-2}
cal mol^{-1}	6.947×10^{-17}	4.184	1	4.336×10^{-5}	3.498×10^{-1}
eV	1.602×10^{-12}	9.649×10^4	2.306×10^4	1	8.066×10^3
cm^{-1}	1.986×10^{-16}	1.196×10	2.859	1.240×10^{-4}	1

One of the units given in column A is equivalent to the value in its row given for each of the several units in column B. For example, to convert 45 000 cal mol^{-1} to eV, multiply by 4.336×10^{-5} to give the result 1.94 eV

Index

255